Table of Contents

	Preface	v
Chapter		Page
1	Understanding Some Principles	1
	Quiz	16
2	Understanding Practical Electricity	19
	Quiz	40
3	The Simple Electrical Circuit	43
	Quiz	55
4	Dc Series Circuits	57
	Quiz	68
5	Dc Parallel Circuits	71
	Quiz	83
6	Combined Series and Parallel Circuits	87
	Quiz	99
7	Understanding Magnetism	101
	Quiz	113
8	Understanding Electromagnetism	115
	Quiz	128
9	Understanding Alternating Current	131
	Quiz	143
10	Working with Resistive Ac Circuits	145
	Quiz	154
11	The Basics of Inductance	157
	Quiz	167
12	Inductive Circuits	169
	Quiz	186
13	Understanding RL Circuits	189
	Quiz	211
14	Understanding Capacitance	215
	Quiz	229
15	Capacitive Circuits	231
	Quiz	247
16	Understanding RC Circuits	251
	Quiz	270

17	Understanding RLC Circuits	275
	Quiz	287
18	Understanding Transformers	289
	Quiz	302
	Glossary	305
	Index	315
	Answers To Quizzes	321

Preface

This book shows you how the principles of electricity apply to the behavior of ac and dc circuits—the building blocks of all electrical and electronic equipment. You will learn about the devices that produce electrical energy and see how electrical energy can be controlled by means of the most commonly used passive devices: switches, resistors, inductors, capacitors, and transformers.

You will find it helpful, but not essential, to have a general familiarity with the basic principles of electricity and electronics. A companion volume in this series, *Understanding Electricity and Electronics Principles*, can provide the desirable background. However, the only essential prerequisites for learning about ac and dc circuits are a familiarity with the basic principles of algebra and the terminology of trigonometry and, most of all, a desire to learn.

Regarding the algebra, you only need to know how to solve equations by substituting specific numerical values for literal terms. If you have access to an engineering or scientific electronic calculator, you do not have to understand the essential meaning of the expressions in order to determine their numerical solutions. But of course, the more you know about mathematics, the easier it is to understand and use the principles described in this book.

Like other books in this series, this book builds understanding step by step. Each sentence presents, supports, or reviews a thought that is important to your understanding of the subject. When the text refers you to a diagram, stop and study the illustration so that you get a mental picture of its general content.

Chapters conclude with a summary of what you have learned in that chapter, a list of new and important terms and a quiz. Study the summary of what you have learned very carefully. None of ideas in the summary should seem new or different from what you recall. If you are uncertain about the meaning of any of the key words, review the relevant text and consult the Glossary. The quiz tests your understanding of important ideas and mathematical procedures. Only one choice is correct, but many of the other choices are based on common misunderstandings of the topic. You can check your responses by consulting the list of answers in the back of the book.

This book has been carefully designed to make your learning process as simple and meaningful as possible. the necessary effort, you can translate the material into ideas that you can remember and use.

D.L.H.

Understanding Some Principles

ABOUT THIS CHAPTER

This chapter introduces you to the basic principles that underlie the effects of electricity. You will learn about molecules, atoms, and ions, and you will be able to visualize how they contribute to electrical effects. You will also learn about conductors, insulators, and the principles of static electricity.

WHAT IS ELECTRICITY?

Electricity is voltage and current. Voltage is electrical pressure, and current is the flow of charged particles called "electrons."

"Voltage" is an excess of electrons at one place compared to another. "Current" is the electron flow resulting from nature's attempt to balance the number of electrons in each place.

Figure 1-1 shows two water tanks which are connected at the bottoms with a pipe. Tank 1 is nearly full of water; Tank 2 is nearly empty. You could say that Tank 1 has an excess amount of water compared to Tank 2. It is said that nature abhors an imbalance. The water in both tanks exerts a downward force, or pressure, but water flows from Tank 1 to Tank 2 because the larger amount of water exerts a much greater force. To balance the two forces, nature allows the water to run from Tank 1 to Tank 2 until the water levels are the same.

> Voltage is electrical pressure that causes current to flow. Current is the flow of electrons.

**Figure 1-1.
Difference in Forces between Two Tanks of Water**

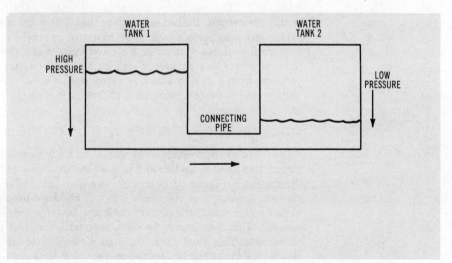

1 UNDERSTANDING SOME PRINCIPLES

Voltage Source

A voltage source represents an imbalance in electron pressure between two points. As long as that imbalance of forces is maintained, current (as electrons) flows from the point of higher electron pressure to the point of lower pressure.

The point of higher electron pressure is called the negative (−) point, and the point of lower pressure is called the positive (+) point. Electrons flow from the negative point, through a suitable path, to the positive point. All voltage sources have these two "polarities."

Figure 1-2 shows current flowing from the negative terminal of a voltage source (a battery), through a load (lamp), and back to the positive terminal of the source. Current can flow through the circuit as long as the battery can maintain a difference in electron pressure between its two terminals.

Electrons flow from the negative terminal of a voltage source, through a suitable path, and return to the positive terminal of the voltage source. The negative and positive terminals represent the two polarities of the source.

**Figure 1-2.
Voltage and Current**

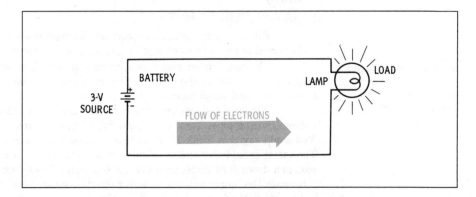

Various kinds of useful voltage sources use some form of energy to maintain a difference in electron pressure between their negative and positive terminals. Batteries maintain that difference in pressure for reasonably long periods of time by means of internal chemical reactions. Electrical generators maintain a desired amount of difference in pressure for indefinitely long periods of time by means of mechanical energy—water pressure from a dam or steam pressure from a boiler. Other kinds of voltage-source devices maintain a difference in electron pressure by heat and light energy.

MOLECULES

To visualize charged particles and their movement through a circuit, you need to understand how all matter is put together. Using the illustration in *Figure 1-3* as a guide, consider the smallest grain of salt you can see. Assume that you break it in half and then break one of the halves in half and continue the process until you have the smallest piece of salt possible. This piece cannot be seen, even with the most powerful microscope. This breakdown to the smallest piece of salt would require several million of the half-by-half steps.

UNDERSTANDING SOME PRINCIPLES

Figure 1-3.
Dividing a Grain of Salt

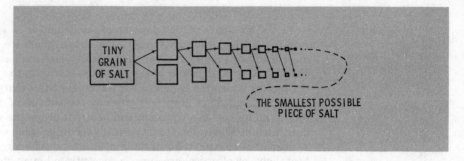

The smallest piece of salt is a single molecule of salt. A "molecule" is the smallest possible particle of a material that retains the essential chemical and physical characteristics of that material. Break down that molecule of salt any farther and you wouldn't have anything recognizable as salt.

What would you have if you did break down a molecule of salt into smaller particles? Molecules are made up of basic elements. A molecule of salt is made up of two elements—one part chlorine and one part sodium. Those elements are chemically bonded together as shown in *Figure 1-4*.

Figure 1-4.
Molecule of Salt

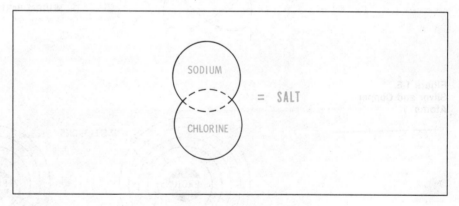

ATOMS

The elements that are bonded together to form a molecule are called atoms. An "atom" is the smallest portion of an element that exhibits all the properties of the element. An "element" is one of a class of substances that cannot be separated into substances of other kinds. There are currently more than 120 known elements in the universe.

The atom, itself, is made up of atomic particles called protons, electrons, and neutrons. There are even smaller particles, called subatomic particles, but they do not play a direct role in the behavior of electricity.

1 UNDERSTANDING SOME PRINCIPLES

All matter is made up of molecules, which comprise atoms of the basic chemical elements. Atoms are made up of electrons, which have a negative electrical charge; protons, which have a positive electrical charge; and neutrons, which have no electrical charge.

Atomic Particles

The atomic-particle makeup of the helium atom is depicted in *Figure 1-5*. The helium atom has two electrons which are in orbit around a nucleus of two protons and two neutrons. An electron has a negative charge that is equal to but of the opposite polarity to the positive charge of the proton. A neutron is neutral—it has no electrical charge.

Atoms differ from each other by the number of electrons in their orbits and by how many protons and neutrons are in the nucleus. As illustrated in *Figure 1-6*, atoms normally have the same number of protons as they have electrons. But if the proper amount of energy, in the form of heat, light, or electrical pressure, is concentrated in an atom, it can cause the atom to give up or take on electrons.

**Figure 1-5.
Helium Atom**

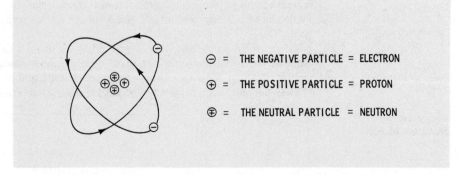

**Figure 1-6.
Silver and Copper Atoms**

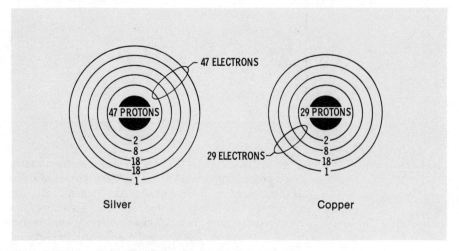

Electrons that are released from their parent atoms are called free electrons.

Free Electrons

Counting outward from the nucleus, the greatest number of electrons that can exist in the first and second orbital zones are 2 and 8, respectively. Subsequent orbital zones may be filled by 8, 18, and 32 electrons. When the atom's outer orbital zone is not completely filled, the

UNDERSTANDING SOME PRINCIPLES

atom is able to release some electrons when voltage or other forms of energy are applied. Electrons that are released from the outer orbital zones of an atom are called "free electrons."

Where do electrons go when they are freed from the parent atom? They usually jump directly to another atom, filling a position in an outer orbital zone previously vacated by another electron.

Silver and copper atoms each have only 1 electron in the outer orbital zone. Less energy is needed to move these electrons to a nearby atom than that required if their outer orbital zones were completely filled.

Inert Elements

Some elements have all of their orbital zones completely filled with electrons. These elements are called "inert" because they neither give up nor accept electrons from another atom. The neon atom is a good example of an inert element. As shown in *Figure 1-7a*, its outer orbital zone contains 8 electrons. Thus, it is completely filled. By contrast, the fluorine atom shown in *Figure 1-7b* has only 7 electrons in its outer orbit, enabling it to accept free electrons. This makes fluorine an "active" element rather than an inert element. Because the neon atom has 10 electrons in orbit around its nucleus containing 10 protons, it is given the atomic number of 10.

**Figure 1-7.
Inert and Active Elements**

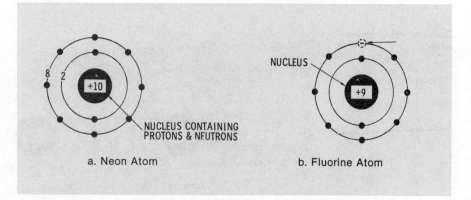

a. Neon Atom   b. Fluorine Atom

Ions

An atom that has more or fewer electrons than normal is called an ion. Positive ions have a shortage of electrons. Negative ions have an excess of electrons.

If enough force or energy is applied to an atom, it is possible to add or take away an electron or two. When this happens, the atom has an unbalanced electrical charge—an unequal number of electrons and protons. This imbalance causes the atom to have either a negative or positive charge. When an atom is thus charged, it becomes an "ion." A negatively charged ion has an excess of electrons and is called a negative ion. A positively charged ion has a shortage of electrons (an excess of protons) and is called a positive ion. See the examples in *Figure 1-8*.

**Figure 1-8.
Ions**

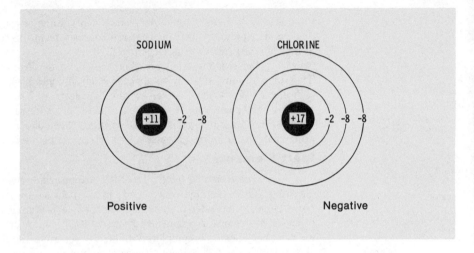

We saw earlier how sodium and chlorine atoms are bonded to form common table salt. Sodium is a positive ion and chlorine is a negative ion. As illustrated in *Figure 1-9*, the two share an electron in their respective outer orbits and are thus electrically bonded to make up a molecule of salt.

**Figure 1-9.
Sodium and Chlorine
Atoms**

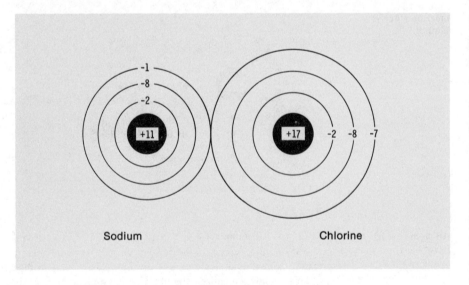

This sharing process establishes a stable combination that is difficult to break. The bond can be broken, however, if sufficient electrical energy is applied to the molecule. In the case of table salt, applying sufficient electrical energy rips apart the bond, producing free ions of sodium and chlorine.

UNDERSTANDING SOME PRINCIPLES

ENERGY PRODUCES ELECTRON FLOW

Energy must be applied to the atoms of a material in order to cause electrical movement. For this electrical movement to be of real value, the energy must be applied in such a manner as to move a relatively large number of electrons—from one atom to the next—in the desired direction.

Heat and light energy can cause electrons to move through a material, but electrical energy is of greatest importance in electronics. Electrical pressure, in the form of a voltage source, is an excessive accumulation of electrons or negatively charged ions in one area with respect to another area.

A battery is one example of a voltage source. It has an excess of negative charges on one terminal (the negative) and a lack of negative charges on a second terminal (the positive).

When the two terminals of a battery are connected to an electrical device through wires, a complete path from one terminal to the other is formed. The negative terminal gives up negative charges that pass through the other elements and return to the positive terminal.

The process continues until the charges on the two battery terminals are equal. Normally a chemical reaction within the battery maintains the imbalance in electrical charges between the two terminals. But even the best batteries eventually run down, and the charges on the terminals approach the same level.

Many processes in electricity are described in terms of electrical charges. Not only will you be concerned with the amount of electrical charge, you will also be concerned with the polarity of the charge.

Conductors and Insulators

A "conductor" is a free-electron material that serves as a path for electron flow. Some materials are better conductors than others. The availability of more free electrons in silver and copper, for example, makes them better conductors than iron and steel. *Figure 1-10* shows some examples of conductors.

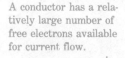

A conductor has a relatively large number of free electrons available for current flow.

Figure 1-10. Conductors

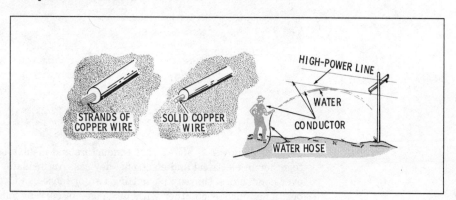

UNDERSTANDING ELECTRICITY AND ELECTRONICS CIRCUITS

1 UNDERSTANDING SOME PRINCIPLES

An insulator has few electrons, if any, available for current flow.

An "insulator" is a material through which electrons do not flow easily. Glass, porcelain, and most plastics are examples of electrical insulators. Such materials have few, if any, available free electrons. Some practical examples of insulator materials are shown in *Figure 1-11*.

**Figure 1-11.
Insulators**

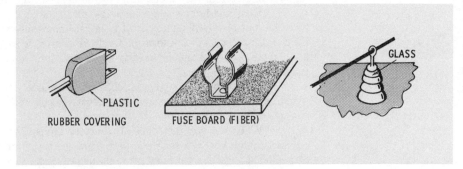

Table 1-1 lists further examples of materials that are generally regarded as good conductors and good insulators.

**Table 1-1.
Common Conductors
and Insulators**

Conductors	Insulators
silver	glass
gold	porcelain
copper	plastic (most)
lead	rubber
tin	mica
brass	ice and snow (pure)
aluminum	nylon
bronze	bakelite
nickel	paper
iron and steel	wood
cadmium	paraffin
graphite	quartz
mercury	dry air

Figure 1-12 shows that conductors and insulators are often used together in electrical and electronic devices. An insulator may be a covering over conductors, thereby permitting two or more conductors to be positioned near each other without danger of current following an undesired path between them. In other instances, an insulator is used to maintain an adequate amount of space between two or more conductors.

UNDERSTANDING SOME PRINCIPLES

Figure 1-12. Conductors and Insulators Used Together

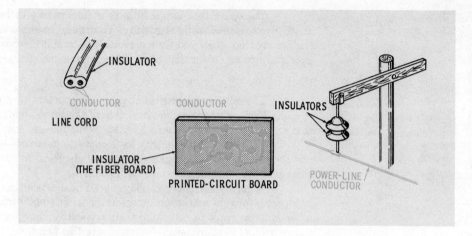

To be effective, the insulator must present a path of high opposition to current flow. Any insulator, however, will become a conductor of electric current if it is subjected to a sufficiently high voltage. If the voltage applied to an insulating material is high enough, the forces rip electrons from the outer atomic orbital zones and create a flow of current, thus "breaking down" the insulator.

STATIC ELECTRICITY

"Static electricity" is the product of a difference in electrical pressure between two materials as a result of frictional forces.

Lightning is an example of static electricity. It is the result of an accumulation of electrical charges caused by friction between clouds, the atmosphere, and the earth, as illustrated in *Figure 1-13*. When the charge becomes great enough, the insulating air breaks down. The resulting discharge—nature's way of balancing the electrical forces—produces the familiar flash of lightning.

Figure 1-13. Charged Clouds Cause Lightning

Cloud collects charge from dry air.

1 UNDERSTANDING SOME PRINCIPLES

The spark that jumps from your fingers to a metal object is another example of static electricity. Your body becomes charged by the friction created when you walk across a rug, and it discharges the energy as a spark to an object that has a different amount of charge.

Electrical Charges

The notion of electrical charges is important to the understanding of many kinds of electrical devices. One of the most important principles is that unlike charges attract one another, while like charges repel. Unlike charges are electrical forces that have opposite polarities—one is negative and the other is positive. Like charges are electrical forces that have the same polarity—both negative or both positive.

A negatively charged body placed near a positively charged body produces a force of attraction between them. One body tries to contact the other in an attempt to neutralize their respective charges. *Figure 1-14* shows a simple experiment to demonstrate this principle. Inflate a balloon and rub it on your hair. (Make certain your hair is clean and dry.) Then hold the balloon near a wall. The balloon clings to the wall because of the difference in charges between those two objects. In this instance, the balloon has a high negative charge compared to the wall.

> Bodies having unlike (opposite) static charges attract one another. Bodies having like (both positive or both negative) charges repel one another.

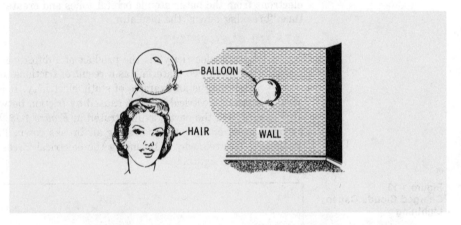

**Figure 1-14.
Balloon with Static Charge**

The same effects can be seen in the pith ball demonstration shown in *Figure 1-15*. If a pith ball is given a negative charge and is then allowed to come in contact with a neutral pith ball, both end up with the same charge and they subsequently repel one another.

**Figure 1-15.
Interaction of Pith Balls**

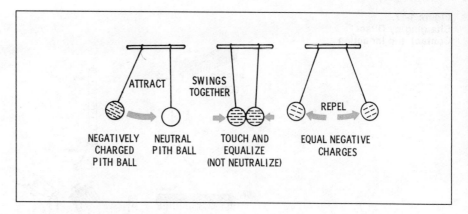

You can perform another experiment, using an electroscope, to demonstrate the attraction and repulsion between static charges. You need an electroscope, such as the homemade version shown in *Figure 1-16*.

**Figure 1-16.
Homemade
Electroscope**

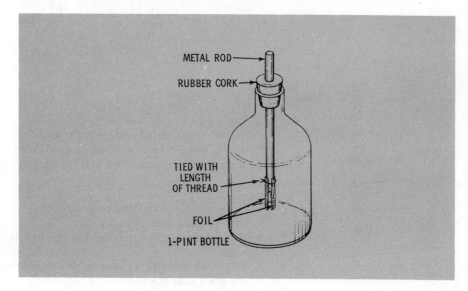

This project requires the following materials.
1. One clear, 1-pint glass bottle and a cork to fit it
2. One metal rod, approximately ¼ inch in diameter
3. Two thin pieces of aluminum foil, measuring about 1 inch × ¼ inch (Foil that has been carefully removed from a chewing gum wrapper is quite suitable for the purpose.)

You can perform two tests with your homemade electroscope, as illustrated in *Figure 1-17*.

1 UNDERSTANDING SOME PRINCIPLES

**Figure 1-17.
Charging by Direct
Contact and Induction**

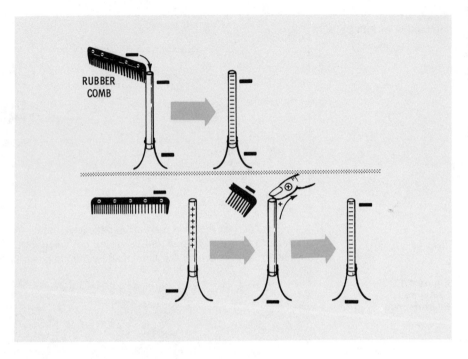

Test 1

1. Rub a plastic comb with fur.
2. Touch the metal rod with your finger.
3. Touch the metal rod with the charged comb.
4. Note that the pieces of foil repel one another.

Test 2

1. Touch the metal rod with your finger.
2. Rub a plastic comb with fur.
3. Bring the comb near the rod, but do not touch the rod with the comb.
4. Touch the rod with a finger on your other hand while still holding the comb near the rod.
5. Remove your finger from the rod.
6. Remove the comb from the vicinity of the rod.
7. Note that the pieces of foil repel one another.

Static charges can be transferred from one body by direct electrical connection or at a distance by the principle of induction.

These tests demonstrate two different ways to transfer electrical charges from one body to another. Test 1 uses a direct electrical contact. Test 2 shows that it is possible to transfer a charge at a distance—by means of a principle called "induction."

UNDERSTANDING SOME PRINCIPLES

You can visualize the effects of charged bodies on one another and on neutral bodies. First, what is the effect called that causes the pith balls and foils to move? The answer is force. What determines the amount and direction of the force? The answer is the amount and polarity of the charge.

Force of Charge

The greater the amount of charge on the two bodies, the greater the amount of force between them. The direction of force depends on the polarity of the charges on the bodies: like charges force them apart, unlike forces tend to pull them together.

Does the distance between two charged objects have any effect on the amount of force between them? Yes. The greater the distance, the less effect one body has on the other. Or, as shown in *Figure 1-18*, the less the distance, the greater the force.

There are many methods used to represent the force between two charged bodies. These methods attempt to make the idea of force between the two bodies easier to understand or visualize. The diagram in *Figure 1-19* emphasizes the difference in electrical charges between two plates that are separated by some distance. The force is equally distributed throughout the space between the plates. The lines drawn between the plates indicate the presence of electrostatic lines of force.

The forces of attraction or repulsion between two electrically charged bodies increase with the amount of charge and decrease with the distance between them.

Figure 1-18.
Less Distance—More Force

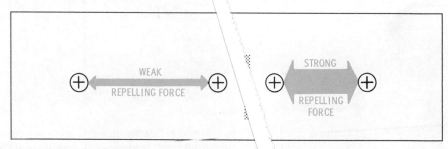

Figure 1-19.
Force Exerted by Two Charged Plates

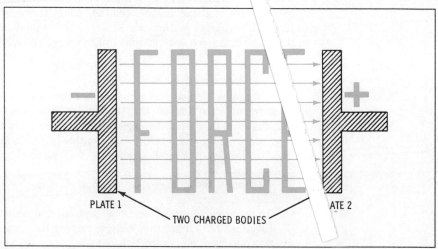

Electrostatic lines of force are not all straight lines. *Figure 1-20* shows that most of the lines of force are curved.

Figure 1-20. Electrostatic Fields between Particles

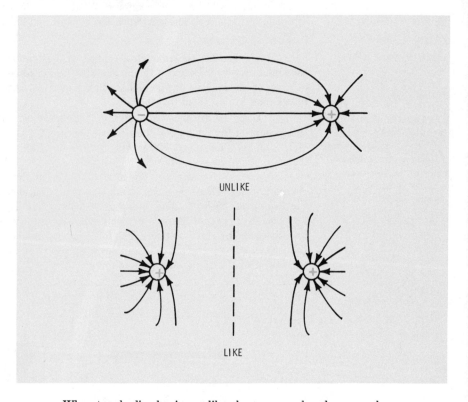

When two bodies having unlike charges are placed near each other, their electrostatic fields attract one another and merge to form a mutual force of attraction between the bodies. However, when bodies having like charges are brought close to one another, their fields repel to form a mutual force of repulsion. Unlike charges attract, and like charges repel.

WHAT HAVE WE LEARNED?

1. Voltage is the electrical pressure that causes current to flow.
2. Current is the flow of charged particles, called electrons.
3. Electrons flow from the negative (−) terminal of a voltage source, through a suitable path, and return to the positive (+) terminal of the voltage source.
4. A molecule is the smallest particle of a material that retains the qualities of that material.
5. An atom is the smallest portion of an element that exhibits all the properties of the element.
6. Atoms are made up of electrons, protons, and neutrons.
7. Electrons have a negative charge, protons have a positive charge, and neutrons have no charge.

UNDERSTANDING SOME PRINCIPLES

8. Atoms normally have equal numbers of electrons and protons.
9. Free electrons have been released from the outer orbits of their parent atoms.
10. Applying a sufficient amount of energy to an atom causes it to release electrons from its outer orbit.
11. An ion has either too many electrons (a negative ion) or too few electrons (a positive ion) in the outer orbits of its atoms.
12. A chemical reaction maintains the imbalance in electrical charges between the negative and positive terminals of a battery.
13. A conductor is a free-electron material that can serve as a path for current flow.
14. An insulator is a material that offers few free electrons. It is a poor conductor.
15. Static electricity is created by frictional energy between two unlike materials.
16. Lightning is the result of static electricity generated by friction between clouds, the atmosphere, and the earth.
17. Like electrical charges repel; unlike charges attract one another.
18. Electrical charges can be passed either by direct contact or through a distance by induction.
19. Distance affects the amount of force that causes charged bodies to attract or repel one another: the greater the distance between the bodies, the smaller the amount of force.

KEY WORDS

Active element	Induction
Atom	Inert element
Battery	Insulator
Cell	Ion
Conductor	Molecule
Current	Neutron
Electron	Proton
Electrostatic lines of force	Static electricity
Element	Voltage
Free electron	

1 UNDERSTANDING SOME PRINCIPLES

Quiz for Chapter 1

1. What is voltage?
   a. The smallest electrically charged particle.
   b. The flow of electrons.
   c. The force that causes electrons to flow.
   d. The number of free electrons in an atom.

2. What is current?
   a. The flow of charged particles through a material.
   b. The force that causes electrons to flow.
   c. An imbalance of electrical charges.
   d. The number of free electrons in an atom.

3. What is the smallest number of polarities required for current flow?
   a. One—a negative terminal.
   b. Two—a negative terminal and a positive terminal.
   c. Three—negative, positive, and neutral terminals.

4. What form of energy is used to maintain an imbalance of charges between the terminals of a battery?
   a. Mechanical energy.
   b. Frictional energy.
   c. Chemical energy.
   d. Heat energy.

5. What form of energy is used for creating static electricity?
   a. Heat energy.
   b. Frictional energy.
   c. Chemical energy.
   d. Light energy.

6. What is a molecule?
   a. The smallest particle of a substance that retains the properties of that substance.
   b. A particle that has an excessive number of electrons in its outer orbit.
   c. A particle that has a shortage of electrons in its outer orbit.
   d. A particle that flows whenever there is an imbalance of electrical charges between two points.

7. What are the main parts of an atom?
   a. Electrons, protons, and ions.
   b. Electrons only.
   c. Electrons, ions, and neutrons.
   d. Electrons, protons, and neutrons.

8. What is the difference between the kinds of charges for an electron and a proton?
   a. An electron has a positive charge and a proton has a negative charge.
   b. An electron has a positive charge and a proton has no charge.
   c. An electron has a negative charge and a proton has a positive charge.
   d. An electron has a negative charge and a proton has no charge.

9. What is a free electron?
   a. An electron that has been freed from its parent atom.
   b. An electron that has no electrical charge.
   c. An electron that has twice the amount of charge of a normal electron.
   d. An electron that costs nothing.

UNDERSTANDING SOME PRINCIPLES

10. What is a negative ion?
    a. An atom that contains more protons than neutrons.
    b. An atom that contains more electrons than neutrons.
    c. An atom that contains more electrons than protons.
    d. An atom that contains more protons than electrons.

11. What is the difference between a conductor and an insulator?
    a. A conductor allows current to flow much easier than an insulator does.
    b. An insulator allows current to flow much easier than a conductor does.
    c. An insulator has an excessive number of free electrons, while a conductor has a few free electrons or none at all.
    d. The difference between conductors and insulators depends on how they are used.

12. Which one of the following sets of conditions creates the greatest amount of electrical attraction between two charged bodies?
    a. Both bodies have the same amount of positive charge.
    b. Both bodies have the same amount of negative charge.
    c. Both bodies have the same amount, but opposite, charges.

Understanding Practical Electricity

ABOUT THIS CHAPTER

In this chapter, you will learn to identify and describe five methods for developing electricity that can serve useful purposes. You will determine the effects of voltage and resistance on current flow and will become familiar with some practical devices.

CURRENTS

Electric current is the movement of electrical charges from one location to another through a conductor. It can also be the movement of charged particles through a battery or any other electrical component. Even lightning is the result of the movement of charges from one location to another mainly ion movement, with positive ions of gas moving in one direction and negative ions in the other.

Current Flow Theory

Charges in motion, then, are actually the movement of either free electrons or ions from one location to another. This constitutes current flow. Some people describe current flow as the movement of positive charges. This is called "conventional current flow." From this viewpoint, charges are said to move from a point of positive charge to one of less positive, or negative, charge. See the example of conventional current flow shown in *Figure 2-1*.

> The theory of conventional current flow assumes that current is made up of positively charged particles flowing from a point of higher positive charge to one having a relatively lower positive, or negative, charge.

**Figure 2-1.
Conventional Current Flow**

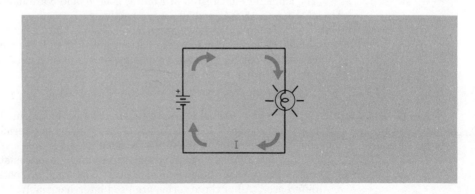

Electron flow theory assumes that current is made up of electrons flowing from a point of higher negative charge to one having a relatively lower negative, or positive, charge.

Other people prefer to think of current as the movement of negative charges. This is the "electron flow" theory. As shown in *Figure 2-2*, electron current is said to flow from the most negative point to a less negative, or positive, point. The electron flow theory is the one favored by most electronics technicians, so it is the viewpoint used in this book.

**Figure 2-2.
Electron Current Flow**

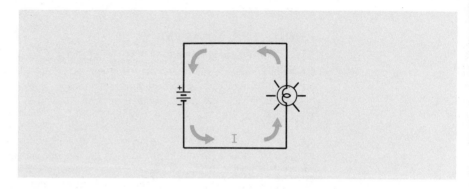

Amperes

The ampere is the unit of measurement for the rate of current flow.

Just as there are units of measurement for height, weight, time, force, and so on, there is also a unit of measurement for current. That unit is the "ampere" (abbreviated A). An ampere is a measure of the rate of flow of electrons past a point in one second (about 6,250,000,000,000,000,000 per second!)

An ampere can be subdivided into smaller units. A milliampere (abbreviated mA), for example, is one-thousandth of an ampere. A microampere (μA) is one-millionth of an ampere.

An ammeter is an instrument used for measuring the amount of current flow.

An ammeter is a device that is used to measure current flow. An ammeter can be scaled in amperes, milliamperes, and microamperes. Some ammeters also have provisions for indicating the direction of current flow. The ammeters illustrated in *Figures 2-3a, b,* and *c* are indicating currents of $+4$ A, -15 mA, and 70 μA, respectively.

Figure 2-4 shows an ammeter inserted into a common kind of lamp circuit. Electron flow from the battery goes through the ammeter and causes the pointer to move an amount that is proportional to the amount of current flowing in the circuit.

VOLTAGE

Voltage is the electrical force that causes current to flow.

The electrical pressure required to move current through a circuit is called "voltage." It is an excess of electrons on one terminal of a voltage source with respect to the other terminal.

As a force, voltage is sometimes called "electromotive force," or emf. Voltage is also known as electrical "potential" or "potential difference." All of these terms are used interchangeably.

UNDERSTANDING PRACTICAL ELECTRICITY

Figure 2-3. Ammeters

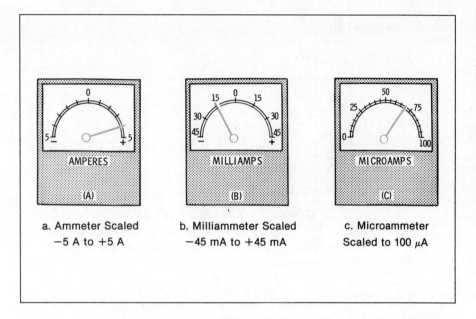

a. Ammeter Scaled −5 A to +5 A

b. Milliammeter Scaled −45 mA to +45 mA

c. Microammeter Scaled to 100 μA

Figure 2-4. Ammeter Inserted into Circuit

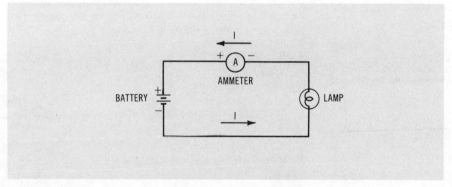

Volts

The volt is the unit of measurement for voltage.

The standard unit of measurement for electrical pressure is the volt (V). Large values of voltage can be expressed in terms of kilovolts (kV), a thousand volts. Smaller amounts of voltage can be expressed as millivolts (mV), which is one-thousandth of a volt, and microvolts (μV), which is one-millionth of a volt.

A voltmeter is the instrument used for measuring voltage.

A voltmeter is a device that is used to measure voltage. *Figure 2-5* shows a voltmeter connected across the battery in a lamp circuit. The potential across the terminals of the battery causes a small amount of current to flow through the voltmeter. That bit of current, in turn, causes the pointer to move an amount that is proportional to the amount of voltage applied to the meter.

**Figure 2-5.
Voltmeter Connected to Circuit**

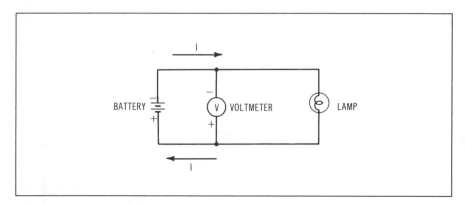

COMMON VALUES FOR CURRENT AND VOLTAGE

Common household electrical devices represent the use of a wide range of current and voltage values. Most flashlights, for example, operate at voltages of 3 V or 6 V, and their current levels range from a couple of milliamperes to nearly 1 A.

Modern homes are wired for operating appliances at 120 V. Large heating appliances, such as electric clothes dryers and electric ranges, operate from 220-V sources. The current levels for some appliances range from a few milliamperes to 20 A. *Figure 2-6* shows the voltage and current rating for several additional household devices.

**Figure 2-6.
Voltage and Current for Household Appliances**

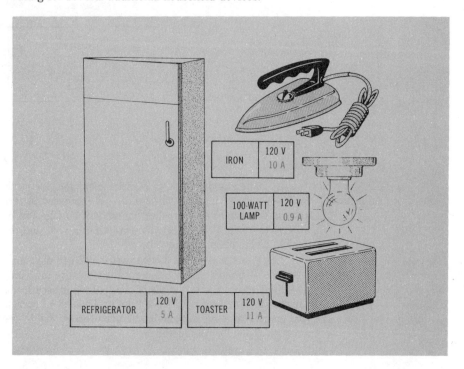

UNDERSTANDING PRACTICAL ELECTRICITY

Most modern automobiles use a 12-V electrical system. Current levels are lowest for operating the car's clock, while the highest—as much as 400 A—is required for operating the starter motor.

SOURCES OF VOLTAGE

All sources of voltage use some mechanism for converting one kind of energy into electrical energy. There are many ways to accomplish this task, but there are just a few fundamental ways. All other mechanisms for generating voltage are variations of the five shown in *Figure 2-7*.

Figure 2-7.
Voltage Sources

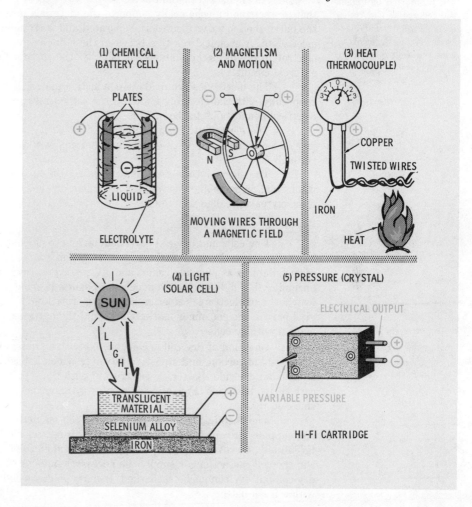

2 UNDERSTANDING PRACTICAL ELECTRICITY

Chemical Voltage Sources

A chemical cell is a device that uses a chemical reaction to produce electrical energy—voltage and current.

Batteries and cells are voltage sources in which chemical action is used to develop a voltage potential. A cell is the basic unit of a chemical voltage source. A battery consists of two or more cells put together.

Each of the four most common chemical cells in use today provides a slightly different voltage level. A carbon-zinc cell is considered to be a 1.5-V source, while a mercury cell is rated at 1.4 V. A lead-acid cell is a 2-V source, and a nickel-cadmium cell is rated at 1.2 V.

The basic parts of a chemical cell are its electrodes and electrolyte. The electrodes react chemically with the electrolyte to produce a voltage.

All chemical cells and batteries are covered by a sturdy coating that protects and mechanically supports the materials inside them. Within the outer shell are two electrodes separated and surrounded by an electrolyte. A chemical reaction between the two electrodes and the electrolyte produces the voltage and supplies the current for operating an external circuit.

The electrodes are made of conductive materials that react chemically with the electrolyte and, at the same time, provide suitable electrical connections to the outside world.

The electrolyte is any substance that easily breaks down into ions, reacts chemically with electrodes, and easily allows current to flow through it.

The electrolyte in a cell may be in a solid or liquid form. Cells using solid or paste-like electrolytes are called dry cells. Those using liquid electrolytes are called wet cells.

Dry Cells

Dry cells and dry-cell batteries have become common household items. They are the small and inexpensive power sources for a vast range of such devices as portable radios and TV receivers, toys, portable lamps (including flashlights), electric pencil sharpeners, hearing aids, digital watches, and electronic cameras. Battery-operated devices have become a vital part of the consumer marketplace—and the batteries are most often made up of dry cells.

The carbon-zinc cell is one of the most commonly used dry cells.

One kind of dry cell is constructed with an outside shell of zinc, a center rod of carbon, and a pasty electrolyte which is basically a solution of ammonium chloride. Batteries constructed with these materials are called carbon-zinc batteries. *Figure 2-8* is a cutaway diagram of a typical carbon-zinc cell.

The chemical reaction within a dry cell maintains the voltage level and supplies current to external circuitry for a reasonable length of time. Eventually, however, the electrode materials and electrolyte are used up, and the cell can no longer supply the necessary amount of current. A dead dry cell battery cannot be recharged. It must be discarded and replaced with a fresh battery.

Figure 2-8.
Carbon-Zinc Dry Cell

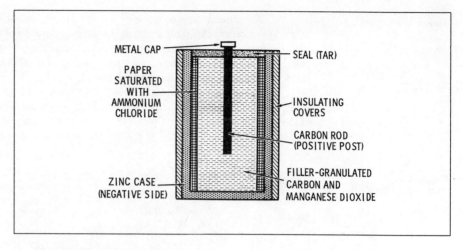

Recall that a fresh carbon-zinc dry cell produces about 1.5 V within its rated level of current. A 1.5-V battery is actually a single cell. Batteries made up of two or more 1.5-V dry cells have voltage ratings that are multiples of 1.5 V. Some common carbon-zinc battery ratings are 6 V, 12 V, 24 V and 48 V. The higher voltage batteries are usually designed for application in high-intensity flashlights and portable camping lights.

Mercury batteries are dry-cell batteries that work on the same principle as carbon-zinc batteries, but the materials are different. The electrodes in a mercury cell are composed of zinc and mercuric oxide. The electrolyte is a mixture of potassium hydroxide and zinc oxide.

As far as overall performance is concerned, each cell in a mercury battery tends to retain its 1.4-V rating until the cells can no longer supply current to an operating circuit. Carbon-zinc cells, on the other hand, tend to show a gradual decline in their output voltage.

Primary-cell batteries, including dry cells such as carbon-zinc and mercury cells, cannot be recharged.

Carbon-zinc and mercury batteries are both examples of "primary-cell" batteries, which cannot be properly recharged. Carbon-zinc batteries can be inserted into a battery charger and "recharged" to some extent. But they are not actually recharging. Rather, they are getting added life by a mild heating effect that stimulates the chemical reaction. The same effect can be achieved by warming them to about 120°F by any other means.

Warning *Mercury batteries should never be placed into a battery charger nor discarded in a fire. Subjected to excessive heat, mercury batteries can explode.*

Wet Cells

Wet-cell batteries use a liquid electrolyte, and they are used in applications that require larger amounts of current than most dry-cell batteries can deliver.

2 UNDERSTANDING PRACTICAL ELECTRICITY

Automobile batteries are lead-acid, wet-cell batteries.

A common type of wet-cell battery is the lead-acid battery, used in autos and motorcycles, and in small, battery-operated vehicles. Among the reasons for lead-acid batteries' popularity for such applications is their reliability, the ability to recharge them quickly and efficiently, and the large amounts of current they can deliver.

Figure 2-9 is a cutaway view of a typical lead-acid battery. The electrodes are made of spongy lead and lead peroxide. The liquid electrolyte is sulfuric acid. Due to the use of this highly corrosive acid, the battery must be constructed of plastic materials that are not attacked by the acid.

**Figure 2-9.
Lead-Acid Wet Cell**

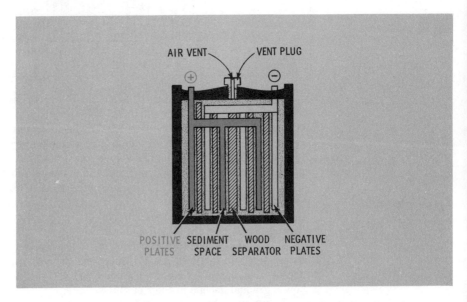

Lead-acid batteries are readily rechargeable. In fact, the alternator (a type of electrical generator) is used in automobiles to recharge the battery. The battery is used in an automobile only when the starter motor is operating or when electrical features are being used and the engine is turned off.

Two special safety precautions must be observed when working with lead-acid batteries. First, beware of getting any of the electrolyte on your skin or in your eyes. Of course, the sulfuric acid will eat holes in your clothing, but that is nothing compared to the skin burns and eye damage that can result from the careless handling of lead-acid batteries.

The second precaution is to keep flame and sparks away from a lead-acid battery that is being recharged. All lead-acid batteries give off hydrogen gas when they are being recharged. Hydrogen gas, mixed with oxygen from the atmosphere, becomes a highly explosive mixture.

UNDERSTANDING PRACTICAL ELECTRICITY

Some newer lead-acid batteries are tightly sealed, and the hydrogen is absorbed by materials inside the case. But older lead-acid batteries and the high-performance versions used in battery-operated vehicles vent the hydrogen gas into the surrounding atmosphere.

Nicad Batteries

Nickel-cadmium cells are wet cells. Unlike lead-acid batteries, however, nicads are tightly and permanently sealed.

Nickel-cadmium (nicad) batteries are wet-cell batteries that are most useful in devices that require frequent recharging but relatively low current demands. Nicad batteries are used in most home small appliances such as portable vacuum sweepers, electric knives, and cordless electric shavers.

The principle of operation of nicads is much the same as that of lead-acid batteries. However, the electrodes in nicad batteries are made of cadmium and nickel hydroxide. The electrolyte is a liquid solution of potassium hydroxide. One of the special advantages of nicad batteries is that they can be permanently sealed. There is no significant build-up of gas pressure when they are being properly recharged, and it is impossible to come into accidental contact with the electrolyte. In other words, the two main precautions for working with lead-acid batteries do not apply to nicad batteries.

Another special feature of nicad batteries is that they tend to deliver their rated voltage and current levels up to the time they require a full recharge. By contrast, the performance of lead-acid batteries shows a gradual decrease in performance, beginning almost from the moment they are no longer being recharged.

One of the disadvantages of nicad cells is that they must not be recharged at a rapid rate. As a rule of thumb, nicads should not be recharged any more than 10 times as fast as they are discharged through normal use. On the other hand, lead-acid batteries can be charged at a much faster rate. In fact, occasionally "deep charging" lead-acid cells at a very high current cleans undesirable materials from the electrode plates.

Secondary-cell batteries, including lead-acid and nicad batteries, can be recharged many times.

Recall that cells that cannot be fully recharged are called primary cells. Cells that can be recharged any number of times, such as lead-acid and nickel-cadmium cells, are called "secondary cells." *Figure 2-10* illustrates two different versions of battery chargers.

The process of recharging a secondary-cell battery does not simply put back the electrons that have been drained from the battery through use. This might appear to be the case, however, because a battery is recharged by connecting it to an electrical device (battery charger) that applies a current and voltage to the terminals of the battery.

The recharging process actually reverses the chemical reaction that took place during the discharging interval. The chemical reaction in primary cells, as such carbon-zinc and mercury cells, cannot be reversed by any means, and therefore it is pointless to attempt recharging primary-cell batteries.

**Figure 2-10.
Battery Chargers**

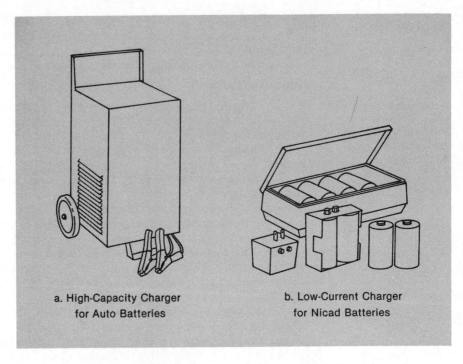

a. High-Capacity Charger
for Auto Batteries

b. Low-Current Charger
for Nicad Batteries

Magnetic Voltage Sources

Motion between a conductor and magnetic field induces a current in the conductor. This principle is the basis of operation for most electrical generators.

Electricity can be generated by moving wire (a conductor) through a magnetic field. Also, electricity can be generated by moving a magnetic field across a number of wires. It makes no difference whether the wire moves, the magnetic field moves, or the wire and field both move. The principle is the same: current will flow in a wire in response to motion between the wire and a magnetic field.

The amount of current that can be generated by using magnetism and mechanical motion depends on several factors. One factor is the strength of the magnetic field. The stronger the magnetic field, the greater the amount of current induced in the wire. Another factor is the speed with which the wire and magnetic field move with respect to one another. As you might suspect, faster motion produces more current flow in the wire. Finally, the length or number of wires is an important factor. The more wire involved in the process, the more current induced in it.

Figure 2-11 shows a technique for determining the direction of electron flow induced in a wire by motion between the wire and a magnetic field. Bearing in mind that lines of magnetic force are directed from the north to the south pole, point your index finger in the direction of that field. Point your thumb in the direction of motion of the wires relative to the magnetic field. If you are using your left hand as shown in the figure, your middle finger indicates the direction of electron flow through the wires.

**Figure 2-11.
Left-Hand Rule**

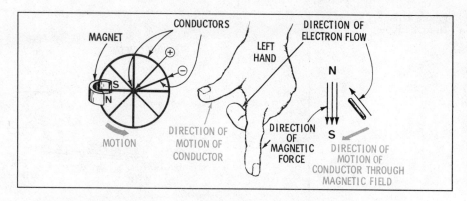

This procedure is known as the "left-hand rule" and, as you have seen, it lets you determine the direction of electron flow induced in the wires. If you use your right hand in the same way, your middle finger points in the direction of conventional current flow.

The fact that motion between a conductor and magnetic field produces electrical energy is the basis for the operation of electrical generators. You have already learned that an alternator (a type of generator) is used in automobiles. Your local electric power company supplies electrical energy that is produced by large generators which get their motion from turbines that are turned by water or steam pressure.

Heat-Generated Voltage

Heat can be used to produce electricity directly by joining two different metals, heating the junction of the metals, and taking the output at the cooler end. Such a device is called a "thermocouple generator." The thermocouple generator illustrated in *Figure 2-12* uses chromel and alumel as the two metals. Other thermocouple metals include alloys of antimony and bismuth.

A thermocouple generator converts heat energy directly into electrical energy.

The thermocouple has many low-power applications. One common application is in an electric thermometer. Thermocouple wire is fixed in a probe that can be inserted into a material or space where the temperature is to be monitored. The junction end of the wires is located near the tip of the probe. A milliammeter is connected to the loose ends of the wire. The thermocouple generates a current that is proportional to the difference in temperature between the tip of the probe and the connections to the milliammeter. The milliammeter, as shown in *Figure 2-13*, is usually scaled in degrees of temperature instead of milliamperes.

Figure 2-12. Thermocouple

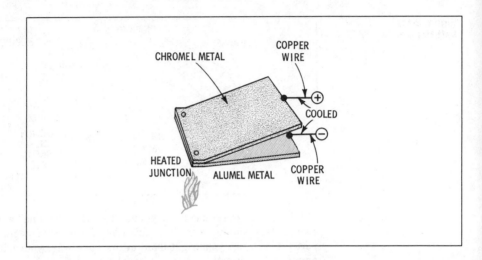

Figure 2-13. Thermocouple in Temperature Probe

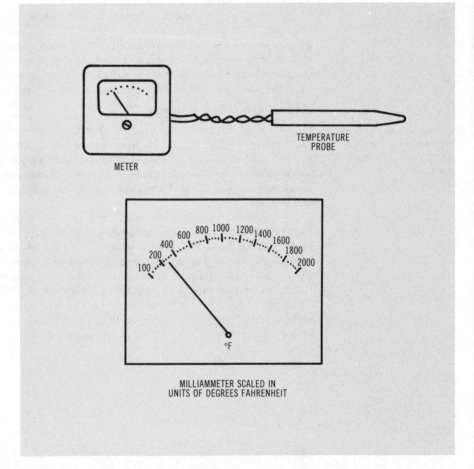

UNDERSTANDING PRACTICAL ELECTRICITY

Light-Generated Voltages

A solar cell converts light energy directly into electrical energy.

A solar cell uses light to produce electricity directly. As indicated in *Figure 2-14*, selenium is one of the materials used in solar cells. The other basic material is iron. Selenium is an element that readily gives up its free electrons under the force of light energy. If iron is used as the backing material, the selenium emits its free electrons into the iron when light strikes the assembly. The iron thus takes on a negative charge relative to the selenium. Wires joined to the two metals form the electrical source terminals.

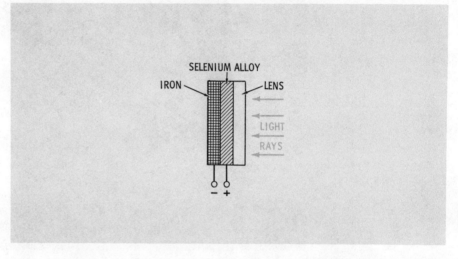

Figure 2-14. Solar Cell

A single solar cell produces little voltage and current. It is possible, however, to interconnect a great many individual solar cells to create a solar panel. Solar panels, directed toward bright sunlight, can produce enough electrical energy to recharge batteries in space satellites. *Figure 2-15* shows a communications satellite that has its solar panels extended to take full advantage of a direct conversion of sunlight to electrical energy.

It is important to realize that most small, light-sensing devices are not solar cells. Photodiodes and phototransistors, for instance, respond to light energy, but do not convert light directly to electrical energy. Rather, these devices change their ability to conduct current in response to light striking their light-sensitive surfaces.

A solar cell is a true light-to-electricity converter.

**Figure 2-15.
Space Satellites Use
Solar-Cell Panels**

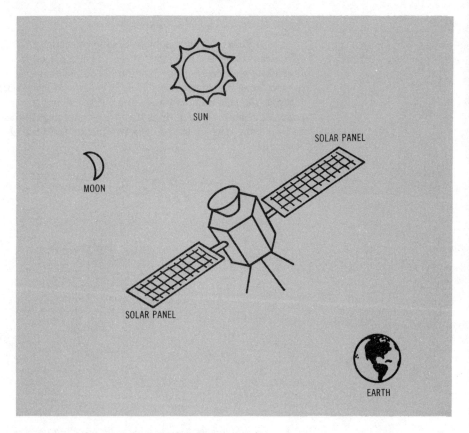

Pressure-Generated Voltages

Certain kinds of crystals, ceramics, and plastics produce electricity when pressure is applied to them. A phonograph cartridge uses this principle. As shown in *Figure 2-16*, the device consists of two metal plates separated by a crystal, and it has a needle that is vibrated by the wavy variations in the groove of the record. These vibrations cause the crystal to be alternately squeezed and released, developing a small, varying voltage across the terminals. Many phonograph cartridges today use ceramic materials in place of the crystalline substance.

Some microphones use the same general principle to convert sound into electrical signals that can be carried through wires to an amplifier and speaker system. Such microphones are specified as crystal or ceramic microphones, depending on the material they use for converting sound directly to electrical energy. *Figure 2-17* shows the main parts of a crystal microphone.

**Figure 2-16.
Phonographic Cartridge**

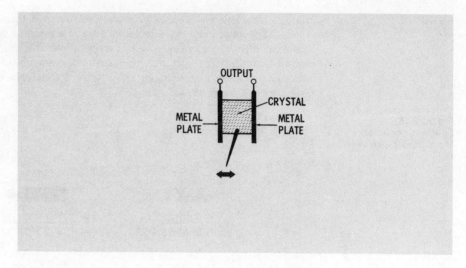

**Figure 2-17.
Crystal Microphone**

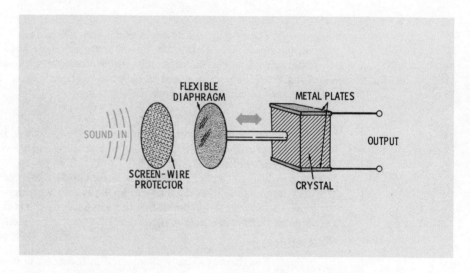

Materials that possess the ability to convert mechanical pressure directly to electrical energy (and vice versa) are called piezoelectric materials.

RESISTANCE

The amount of current flowing in a circuit is determined by how much voltage is applied and how difficult it is for the current to flow. "Resistance" is the opposition to current flow in a circuit.

How easily current will flow through a conductor depends on the availability of free electrons in the conductor. The more free electrons, the less resistance the material offers to current flow. Conductors having

Resistance is the property of a material that tends to resist the flow of current through the material.

relatively few free electrons offer more electrical resistance to current flow.

The length and cross-sectional area of a material have a great influence on its overall resistance to current flow. The longer the path for current flow through a material, the greater the amount of resistance. As indicated in *Figure 2-18*, doubling the length of a material doubles its resistance to current flow.

**Figure 2-18.
Longer Conductor—
More Resistance**

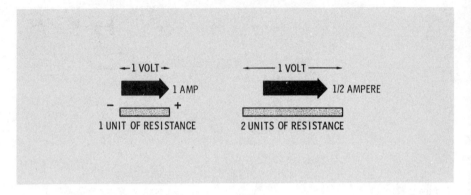

The greater the cross-sectional area is, the lower the resistance will be. *Figure 2-19* shows that doubling the cross-sectional area of a material cuts its resistance in half. Other characteristics, such as the kind of material, temperature, and the kind of electricity (ac or dc), also affect resistance to current flow as well.

Ohms

The ohm is the unit of measurement for electrical resistance.

The standard unit of measurement for resistance is the ohm (Ω). One ohm of resistance limits current to 1 ampere when 1 volt of electrical pressure is applied.

The amount of resistance is designated by the number of ohms. Very large values of resistance can be expressed with the standard prefixes of kilo- and meg-. A kilohm (kΩ) is the same as one thousand ohms. A megohm (MΩ) is one million ohms.

Resistors

Electrical resistance is inserted into circuits in order to control the amount of current flow. Simple devices called resistors are manufactured to have a specific amount of resistance. Resistors of various values are used in every sort of electronic device, from simple portable radio receivers to complex computer systems.

Resistors can be classified as fixed and variable. You can adjust the resistance value of a variable resistor by turning a shaft. You cannot change the resistance value of a fixed resistor.

Resistors are manufactured in many different forms and with many different resistance values. As shown in *Figure 2-20*, resistors can also be classified as fixed or variable. A fixed resistor has a value that cannot be changed. A variable resistor is constructed so that its value can be easily changed. The volume control on radio and television receivers is an example of a variable resistor.

**Figure 2-19.
More Area—Less
Resistance**

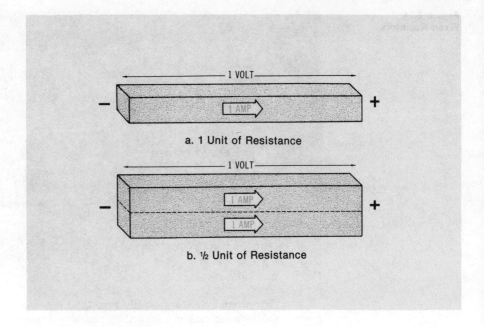

**Figure 2-20.
Resistor Types**

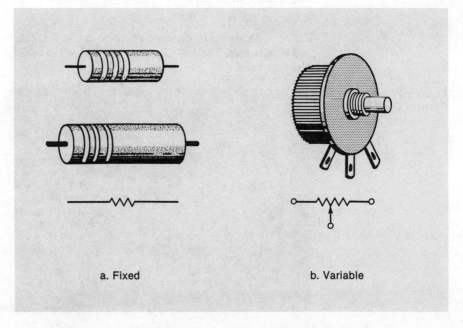

Figure 2-21 shows two common kinds of fixed resistors: one of carbon composition, and one of wire wrapped around an insulating core.

**Figure 2-21.
Fixed Resistors**

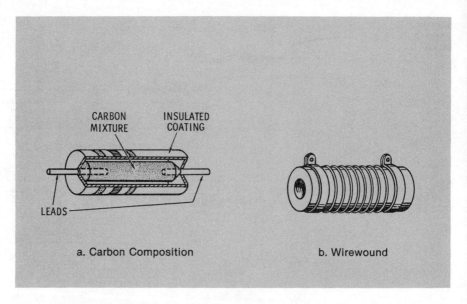

Wirewound resistors usually have their resistance value printed directly on them but carbon composition resistors usually show their specifications by means of color-coded bands painted on them. It is important that you know how to "read" the color coding on such resistors.

Resistor Ratings

As shown in *Figure 2-22*, there are at least three color bands and sometimes four. The one closest to one end of the resistor is regarded as the first band.

**Figure 2-22.
Resistor Color Bands**

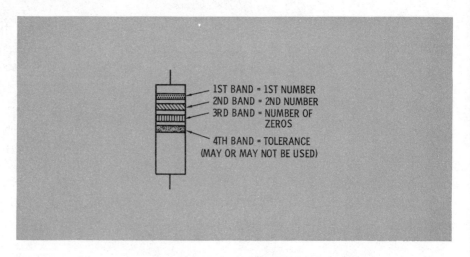

UNDERSTANDING PRACTICAL ELECTRICITY

The first two color bands indicate the first two digits in the resistance value. The third band indicates the number of zeros to follow the two digits. *Table 2-1* shows the ten colors associated with numerals 0 through 9.

Table 2-1.
Resistor Color Codes: Resistance

Color	Value	Color	Value
Black	0	Green	5
Brown	1	Blue	6
Red	2	Violet	7
Orange	3	Gray	8
Yellow	4	White	9

As an example, suppose you are looking at a resistor having this combination of color bands: first band is brown, second band is black, and the third band is orange.

In our example, the first digit in the resistance value is 1 (brown), and the second digit is 0 (black). The number of zeros to follow those digits is 3 (orange). Therefore, the value of the resistor is 10,000 Ω or 10 kΩ.

The fourth color band indicates the tolerance of the resistor. "Tolerance" specifies how much the actual resistance can vary from the indicated value of the resistor. *Table 2-2* shows that a gold fourth color band specifies a tolerance of 5%, a silver fourth band 10%, and no fourth band indicates a tolerance of 20%.

Table 2-2.
Resistor Color Codes: Tolerance

Color	Tolerance
Gold	5%
Silver	10%
No color	20%

Consider that 10% of 1000 Ω is 100 Ω. The actual resistance of a 1000-Ω resistor with a tolerance rating of 10% can vary 100 Ω above and below 1000 Ω. In other words, the actual resistance can be as low as 900 Ω and as high as 1100 Ω.

Resistance is measured with an instrument called an ohmmeter.

WHAT HAVE WE LEARNED?

1. Electric current is the movement of electrical charges.
2. The conventional theory of current flow states that current flows from a positive to a negative source of charges.
3. The electron theory of current flow states that current flows from a negative to a positive source of charges.
4. The ampere is a measure of the rate of electron flow.
5. The basic unit of measurement for current is the ampere (A). Smaller units in common use are the milliampere (1 A = 1000 mA) and the microampere (1 A = 1,000,000 µA).
6. An ammeter is an electrical instrument used for measuring current.
7. Voltage is the electrical force that causes current to flow through a circuit.
8. Voltage is also variously called electromotive force (emf), electrical potential, and potential difference.
9. The basic unit of measurement for voltage is the volt (V). Other units include kilovolt (1 kV = 1000 V), millivolt (1 V = 1000 mV), and microvolt (1 V = 1,000,000 µV).
10. A voltmeter is an electrical device that is used for measuring voltage.
11. A cell is the basic element for producing voltage. A battery is two or more chemical cells connected together to provide more voltage or current than a single cell can.
12. The essential parts of a chemical cell are the electrodes and the electrolyte. A chemical reaction between the electrodes and electrolyte produce a voltage.
13. Dry-cell batteries use a solid or paste-like electrolyte, whereas wet-cell batteries use a liquid electrolyte.
14. Primary-cell batteries cannot be properly recharged, whereas secondary-cell batteries can be fully recharged many times.
15. Lead-acid batteries are wet-cell batteries that use lead compounds for the electrodes and sulfuric acid for the electrolyte.
16. Nickel-cadmium (nicad) cells are wet-cell batteries that use nickel and cadmium for the electrodes and a solution of potassium hydroxide for the electrolyte.
17. Motion between a conductor and a magnetic field generates a current in the conductor. A generator working according to this principle converts mechanical energy into electrical energy.
18. A thermocouple converts thermal energy to electrical energy.
19. A solar cell converts light energy to electrical energy.
20. Piezoelectric materials convert mechanical pressure into electrical energy.
21. Resistance is the opposition to current flow through a circuit.
22. The basic unit of measurement for resistance is the ohm (Ω). Larger values of resistance are expressed in kilohms (1 kΩ = 1,000 Ω) and megohms (1 MΩ = 1,000,000 Ω).
23. An ohmmeter is a device used for measuring resistance.
24. A resistor is a device that is designed to provide a specified amount of resistance in a circuit.
25. The resistance and tolerance values of composition resistors are usually color coded on the body of the resistor.

UNDERSTANDING PRACTICAL ELECTRICITY 2

KEY WORDS

Ammeter
Ampere
Battery
Carbon-Zinc Cell
Cell
Dry-Cell Battery
Electrode
Electrolyte
Electromotive force
Emf
Kilohm
Kilovolt
Lead-Acid Cell
Left-Hand Rule
Megohm
Mercury Cell
Microampere

Microvolt
Milliampere
Millivolt
Nickel-Cadmium Cell
Ohm
Ohmmeter
Piezoelectric Effect
Potential Difference
Primary-Cell
Resistance
Resistor
Secondary-Cell
Solar Cell
Thermocouple
Voltage
Voltmeter
Wet-Cell Battery

Quiz for Chapter 2

1. The direction of conventional current flow is:
   a. from left to right.
   b. from positive to negative.
   c. from negative to positive.

2. The direction of electron current flow is:
   a. from right to left.
   b. from positive to negative.
   c. from negative to positive.

3. The basic unit of measurement for current is the:
   a. watt.
   b. volt.
   c. ampere.
   d. ohm.

4. The basic unit of measurement for electromotive force (emf) is the:
   a. watt.
   b. volt.
   c. ampere.
   d. ohm.

5. The basic unit of measurement for resistance is the:
   a. ampere.
   b. volt.
   c. ohm.
   d. watt.

6. The equivalent of 1 kV is:
   a. $1/1000$ V.
   b. 1000 V.
   c. 1 mV.
   d. 1000 mV.

7. The equivalent of 1 µA is:
   a. $1/10,000$ A.
   b. 1000 A.
   c. one-millionth of an ampere.
   d. one million amperes.

8. The equivalent of 1 megohm is:
   a. 10 kΩ.
   b. 100,000 Ω.
   c. one-millionth of an ohm.
   d. one million ohms.

9. Which one of the following devices is most often used for measuring electrical potential difference?
   a. Voltmeter.
   b. Ammeter.
   c. Ohmmeter.
   d. Wattmeter.

10. Which one of the following devices is most often used for measuring opposition to current flow in a circuit?
    a. Voltmeter.
    b. Ammeter.
    c. Ohmmeter.
    d. Wattmeter.

11. Which one of the following is most likely used for measuring the rate of electron flow?
    a. Voltmeter.
    b. Ammeter.
    c. Ohmmeter.
    d. Wattmeter.

12. A device that converts chemical energy to electrical energy is called a:
    a. thermocouple.
    b. solar cell.
    c. chemical cell.
    d. piezoelectric device.
    e. generator.

13. A battery is a device composed of one or more:
    a. chemical cells.
    b. solar cells.
    c. voltage generators.
    d. piezoelectric cells.
    e. electrical terminals.

UNDERSTANDING PRACTICAL ELECTRICITY

14. A device that converts magnetic/mechanical energy to electrical energy is called a:
    a. thermocouple.
    b. solar cell.
    c. chemical cell.
    d. piezoelectric device.
    e. generator.

15. A device that converts thermal energy to electrical energy is called a:
    a. thermocouple.
    b. solar cell.
    c. chemical cell.
    d. piezoelectric device.
    e. generator.

16. A device that converts light energy to electrical energy is called a:
    a. thermocouple.
    b. solar cell.
    c. photodiode.
    d. phototransistor.
    e. piezoelectric device.

17. A device that converts mechanical pressure to electrical energy is called a:
    a. thermocouple.
    b. solar cell.
    c. photodiode.
    d. nicad cell.
    e. piezoelectric device.

18. What is the primary difference between a wet-cell and a dry-cell battery?
    a. A dry-cell battery usually has only one electrode.
    b. Wet-cell batteries use a liquid electrolyte, while dry-cell batteries do not.
    c. Wet-cell batteries can provide higher voltage ratings.
    d. Wet-cell batteries are usually smaller than dry-cell batteries.

19. What is the primary difference between primary- and secondary-cell batteries?
    a. Primary-cell batteries can be recharged and used many times, whereas secondary-cell batteries cannot be properly recharged.
    b. Secondary-cell batteries can be recharged and used many times, whereas primary-cell batteries cannot be properly recharged.
    c. Primary-cell batteries must be recharged at a much slower rate than secondary-cell batteries.
    d. Primary-cell batteries use chemical cells, whereas secondary-cell batteries use an alternative energy source, such as thermal or magnetic energy, to supplement their outputs.

20. Which one of the following statements applies to common auto batteries?
    a. Auto batteries are primary wet-cell batteries.
    b. Auto batteries are secondary wet-cell batteries.
    c. Auto batteries are primary dry-cell batteries.
    d. Auto batteries are secondary dry-cell batteries.

21. Which one of the following statements applies to common, non-rechargeable flashlight batteries?
    a. Common flashlight batteries are primary wet-cell batteries.
    b. Common flashlight batteries are secondary wet-cell batteries.
    c. Common flashlight batteries are primary dry-cell batteries.

d. Common flashlight batteries are secondary dry-cell batteries.

22. What kind of battery is most likely included in a cordless electric knife?
    a. Zinc-carbon battery.
    b. Mercury-cell battery.
    c. Lead-acid battery.
    d. Nicad battery.

23. A resistor has a rated value of 100 Ω and a tolerance of 10%. What is its minimum and maximum allowable resistance value?
    a. 99 Ω–101 Ω
    b. 90 Ω–110 Ω
    c. 99.9 Ω–100.1 Ω
    d. 98 Ω–102 Ω

24. Determine the resistance value and tolerance of a resistor that has the following color bands: first band is red, second band is red, third band is orange, and the fourth band is gold.
    a. 223 Ω, 5%
    b. 223 Ω, 10%
    c. 22,000 Ω, 5%
    d. 22,000 Ω, 10%
    e. 200,000 Ω, 5%
    f. 200,000 Ω, 10%

The Simple Electrical Circuit

ABOUT THIS CHAPTER

You are now going to learn about basic circuits. You will study the application of Ohm's law and the electrical power relationships that exist between voltage, current, and resistance. Given this information, you will be able to identify basic kinds of circuits, and visualize and describe their characteristics of voltage, current, resistance, and power.

A BASIC CIRCUIT

An electrical circuit is composed of a voltage source and a closed path for current flow from one terminal of the source to the other.

An electrical circuit consists of a closed path through which an electric current flows. The circuit is the route followed by the current as it travels through the conductors from the voltage source, to the load, and back to the source. This route includes the internal path from terminal to terminal within the voltage source itself.

A circuit must form a complete and unbroken loop for current to flow. As indicated in *Figure 3-1*, current flows through a complete circuit in somewhat the same manner as a bicycle chain makes a complete loop between the driving gear (source of energy) and the wheel (load).

The most common sources of dc voltage are batteries and dc generators.

A battery is an example of a direct-current (dc) source of voltage and current. A dc source is one that has one terminal that is always a negative (−) terminal and another that always serves as the positive (+) terminal. Dc generators convert mechanical motion directly into dc voltage and current. *Figure 3-2* shows the schematic symbols for a battery and dc generator.

The current-carrying path between the terminals of a voltage source and its load is the conductor. A conductor can be a wire made of copper or of copper with a thin coating of silver. A conductor can also be the copper plating on a printed-circuit board or the metal that forms part of the structure for a piece of electrical equipment.

SWITCHES

The main purpose of a switch is to turn an electrical device on and off. Switches can also be used to turn current off and on in individual sections of more complex circuits.

3   THE SIMPLE ELECTRICAL CIRCUIT

Figure 3-1.
A Complete Circuit

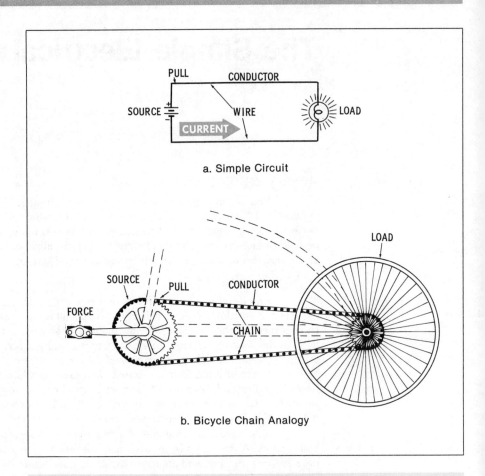

Figure 3-2.
Schematic Symbols, Dc Voltage Sources

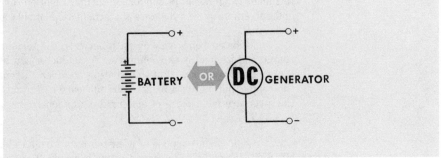

THE SIMPLE ELECTRICAL CIRCUIT

Current flows freely through a closed switch. It cannot flow through an open switch.

A switch that is closed acts as a good conductor, allowing current to flow freely through it. A switch that is open acts as a good insulator, preventing current from flowing through it.

A basic switch is made up of a set of two contacts. One is stationary and the other is movable. The switch is closed when the movable contact makes a firm connection with the stationary contact. The switch is open when the movable contact is physically separated by some distance from the stationary contact.

Knife Switches

Figure 3-3 illustrates the action of the most elementary kind of switch, called a "knife switch." The schematic diagrams show how an open knife switch interrupts the flow of current between a battery and a lamp and how a closed knife switch allows current to flow in the circuit.

Figure 3-3. Knife Switch

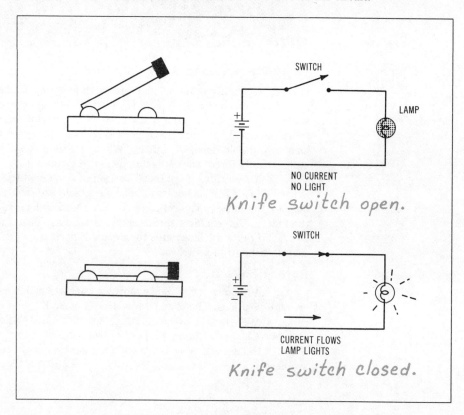

A "toggle switch" is an improved version of a knife switch. Toggle switches use a spring-loaded mechanism to snap the connections open or tightly closed. *Figure 3-4* shows two kinds of toggle switches.

UNDERSTANDING ELECTRICITY AND ELECTRONICS CIRCUITS

3 THE SIMPLE ELECTRICAL CIRCUIT

**Figure 3-4.
Common Toggle Switches**

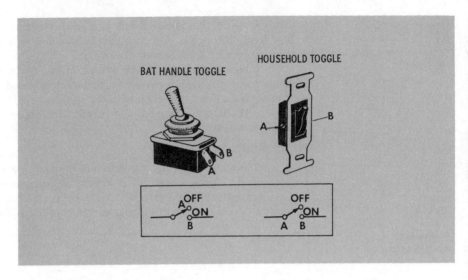

Momentary Pushbutton Switches

Momentary pushbutton switches are classified as normally open and normally closed.

A slightly different kind of switch is spring loaded to favor either its open or its closed state. This is called a "momentary pushbutton switch." The spring in a normally open (N/O) pushbutton switch holds the switch contacts open until you depress it. The contacts remain closed as long as you hold down the button. When you release the button, the spring automatically forces the contacts open once again.

A normally closed (N/C) version of a momentary pushbutton switch is spring loaded in such a way that its contacts are firmly closed until you depress the button. The contacts remain closed as long as you hold down the button. The contacts spring open when you release the button.

Figure 3-5 illustrates the action of normally open and normally closed pushbutton switches.

Rotary Switches

A "rotary switch" has movable contacts which can be set to two or more positions. The channel selector on most TV sets is an example of a rotary switch. Rotary switches are sometimes called "selector switches." *Figure 3-6* shows a simple six-position rotary switch.

In later discussions we will look at other ways to classify electrical switches.

THE SIMPLE ELECTRICAL CIRCUIT

**Figure 3-5.
Momentary Pushbutton
Switches**

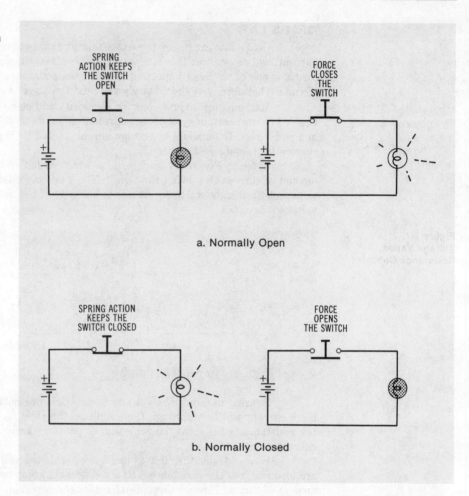

**Figure 3-6.
Six-Position Rotary
Switch**

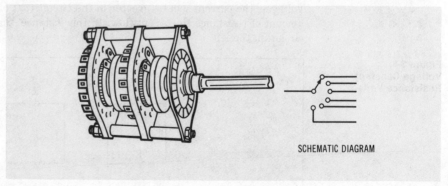

OHM'S LAW

Ohm's law expresses the exact relationship between voltage, current, and resistance. This law, expressed in three simple mathematical forms, is one of the most important laws for understanding and using dc circuits. Therefore, you should study the next few pages very carefully.

Mathematical expressions for electricity and electronics, including Ohm's law, use certain symbols to represent amounts of voltage, current, and resistance. The amount of voltage is represented by the letter E, current by I, and resistance by R.

Refer to *Figure 3-7* to determine which circuit will have the larger amount of current (I) flowing through it. The value of resistance (R) is the same in both instances. However, one is being operated from a higher voltage source (E).

Mathematical equations for electricity and electronics use E for voltage, I for current, and R for resistance.

**Figure 3-7.
Voltage Varied,
Resistance Constant**

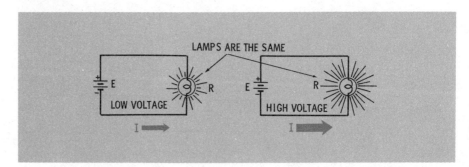

Bearing in mind that voltage is electrical pressure, it follows that more current will flow through the circuit on the right. Assuming that the resistance is the same, larger voltages produce larger amounts of current flow.

Now consider the circuits in *Figure 3-8*. If the voltage levels (E) are identical, but the resistances (R) are different, which has the greater amount of current flowing through it? Under the condition of equal voltages, the current will be greater in the circuit that offers the lesser amount of resistance to current flow. In this instance, that would be the circuit on the left.

**Figure 3-8.
Voltage Constant,
Resistance Varied**

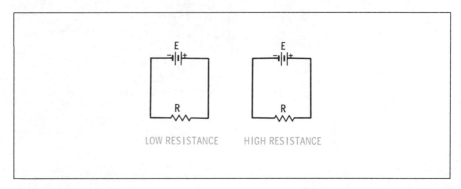

48   UNDERSTANDING ELECTRICITY AND ELECTRONICS CIRCUITS

THE SIMPLE ELECTRICAL CIRCUIT

Solving For Current

Ohm's law not only expresses these relationships in a mathematical form, but it also lets you calculate the exact values. The most fundamental form of Ohm's law is:

$$I = \frac{E}{R}$$

Current is equal to voltage divided by resistance.

Suppose that E = 10 V and R = 5 Ω. How much current will flow under those conditions? Well, current (I) equals E divided by R. So

$$I = \frac{10}{5} = 2 \text{ A.}$$

Solving For Voltage

In the chapters that follow, you will see the need for determining the voltage when current and resistance are known. For example, how much voltage is required to force 2 A of current through a load resistance of 50 Ω? You have already seen that I = E/R. Substituting the known values into that expression leads to this version:

$$2 = \frac{E}{50}$$

The trick is to determine the value of voltage, E. What value, divided by 50 yields 2? You should conclude that E = 100 V. And that is indeed the amount of voltage required for forcing 2 A to flow through a 50-Ω resistance.

By using a mathematical rule, you multiply both sides of the basic equation by R to obtain an equation that shows how to solve for E in a more direct fashion. The result is:

$$E = IR$$

This expression says, voltage is equal to current multiplied by resistance.

Use that form of Ohm's law to calculate the amount of voltage required for forcing a current flow of 10 A through a 12-Ω resistance. Here I = 10 and R = 12. Substituting those known values into the voltage version of Ohm's law: E = 10 × 12, or E = 120 V.

Solving for Resistance

The third form of Ohm's law solves for resistance (R), given the values of voltage (E) and current (I). The mathematical form is:

$$R = \frac{E}{I}$$

This form of Ohm's law says, resistance is equal to voltage divided by current. Having that fact at hand, you should be able to determine how much resistance is required to limit the current from a 100-V source to 0.1 A.

In this case, E = 100 V and I = 0.1 A. Substituting those values into the equation for resistance:

$$R = \frac{100}{0.1} = 1000 \ \Omega$$

Table 3-1 at the end of this chapter includes a summary of Ohm's law equations.

VOLTAGE DROP

Voltage drop is a term that can be misleading. The word "drop" may lead you to believe that an amount of voltage is lost. This is not true, however, because voltage never disappears. Rather, "voltage drop" refers to the manner in which the source voltage is distributed (dropped) throughout a circuit.

All of the source voltage is distributed proportionally across the resistance in a circuit. The voltage drop between any two points in a circuit can be determined by the ratio of the resistance between those points and the total resistance of the circuit. *Figure 3-9*, for example, shows three different resistances in a circuit that is operated from a 6-V source.

Figure 3-9. Voltage Drops

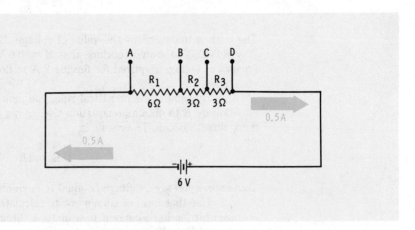

The total resistance in this circuit is 12 Ω. Half that amount of resistance is found between points A and B (resistor R_1). The voltage drop between those points is thus half the source voltage, or 3 V in this instance.

THE SIMPLE ELECTRICAL CIRCUIT

On the other hand, the resistance between points B and C is just 3 Ω, or ¼ the total resistance in the circuit. That means you should find ¼ of the total source voltage between these two points, so the voltage drop is 1.5 V between B and C. The same is true for points C and D.

Voltage drop is sometimes called "IR drop" because the amount of voltage can be determined by one form of Ohm's law: E = IR. If you know the current through a resistor and its resistance, you can calculate the IR drop across it.

ELECTRIC POWER

When voltage forces current through a resistance, heat is generated; the electrical energy in a resistance is converted to heat energy. The rate at which this conversion takes place is called "power." The unit of measurement for electrical power is the "watt" (abbreviated W). A 100-W light bulb, for example, converts electrical energy to heat (and light) at twice the rate of a 50-W light bulb.

Power can be determined by multiplying the applied voltage by the amount of current:

$$P = IE$$

where,

P is the power in watts,
I is the current in amperes,
E is the voltage in volts.

The 120-V dc generator in *Figure 3-10* supplies 0.5 A to a 60-W light bulb. You can confirm that the bulb is indeed dissipating 60 W of power by multiplying the voltage times the current: 120 V × 0.5 A = 60 W.

Power is the rate of conversion of electrical energy to some other form of energy—usually heat. The unit of measurement for electrical power is the watt.

**Figure 3-10.
Power Dissipation in Lamp**

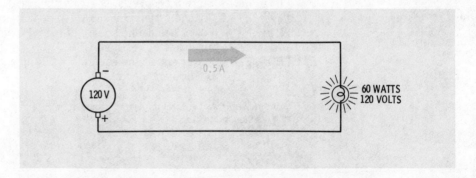

Power Equations

Just as Ohm's law can be expressed in three different mathematical forms, so can the basic power equation. The power equations that use P, I, and E are:

3   THE SIMPLE ELECTRICAL CIRCUIT

$$P = IE$$
$$I = \frac{P}{E}$$
$$E = \frac{P}{I}$$

Now you can answer this sort of question: How much current is required for operating a 100-W light bulb in a 120-V circuit? The question provides values for power (P) and voltage (E), so the appropriate equation is I = P/E. It is the only equation that solves for I, given the values of P and E. In this case, P = 100 and E = 120. Plugging those values into the appropriate equation: I = 100/120 = 0.83 A.

Power can also be determined by two other equations:

$$P = I^2 R$$
$$P = \frac{E^2}{R}$$

where,

$I^2 = I \times I,$
$E^2 = E \times E.$

Tables 3-1 and 3-2 summarize these power equations as well as all versions of Ohm's law.

**Table 3-1.
Summary of Power Equations**

(1)	P = IE	Power equals current multiplied by voltage
(2)	I = P/E	Current equals power divided by voltage
(3)	E = P/I	Voltage equals power divided by current
(4)	P = I²R	Power equals the square of the current (I times I) multiplied by resistance
(5)	P = E²/R	Power equals the square of the voltage (E times E) divided by resistance

P is the power in watts, I is the current in amperes, E is the voltage in volts, and R is the resistance in Ohms.

THE SIMPLE ELECTRICAL CIRCUIT

Table 3-2.
Summary of Ohm's Law

(1) I = E/R   Current equals voltage divided by resistance
(2) E = IR   Voltage equals current multiplied by resistance
(3) R = E/I   Resistance equals voltage divided by current

I is the current in amperes, E is the voltage in volts, and R is the resistance in Ohms.

Because electrical energy is dissipated in the form of heat in a resistance, the power developed in a resistance is considered to be a loss. If 1 A of current causes a voltage drop of 120 V, the power loss (IE) is 120 W. If 2 A of current flows through a 10-Ω resistance, the power loss (I²R) is $2^2 10 = 4 \times 10$, or 40 W. A voltage drop of 10 V across 20 Ω of resistance will dissipate 5 W of power (E²/R).

WHAT HAVE WE LEARNED?

1. An electrical circuit is composed of a voltage source and a closed path for current flow from one terminal of the source to the other.
2. A battery is a common example of a direct-current (dc) voltage source.
3. Current cannot flow through a switch that is in its open position; current flows easily through a switch that is in its closed position.
4. Momentary-pushbutton switches are classified as normally open (N/O) or normally closed (N/C).
5. A rotary switch has a movable contact that can be set to two or more positions.
6. Ohm's law expresses the exact relationship between voltage (V), current (I), and resistance (R). The basic form of Ohm's law is I = E/R, and the variations are E = IR and R = E/I.
7. The voltage dropped across a resistance is sometimes called an IR drop because its value can be determined by multiplying current through the resistance by the value of the resistance.
8. Power is the rate of conversion of electrical energy to some other form of energy—usually heat. The unit of measurement for electrical power is the watt (W).
9. The basic power equation is P = IE, where P is the amount of power in watts, I is the amount of current flowing through the resistance, and E is the voltage dropped across the resistance.

3 THE SIMPLE ELECTRICAL CIRCUIT

KEY WORDS

Conductor
Dc generator
IR drop
Knife switch
Normally closed switch
Normally open switch
Ohm's law
Power
Pushbutton switch
Rotary switch
Selector switch
Toggle switch
Voltage drop
Watt

THE SIMPLE ELECTRICAL CIRCUIT

Quiz for Chapter 3

1. Current can flow through a normally closed pushbutton switch as long as:
   a. you are pressing the button.
   b. you are not pressing the button.
   c. its contacts remain open.

2. Current can flow through a normally open pushbutton switch as long as:
   a. you are pressing the button.
   b. you are not pressing the button.
   c. its contacts remain open.

3. A typical light switch in your home is a:
   a. knife switch.
   b. toggle switch.
   c. momentary pushbutton switch.
   d. rotary switch.

4. Which one of the following equations most directly determines the amount of current flowing through a resistor when you know the value of the resistor and the voltage drop across it?
   a. $E = IR$.
   b. $E = I^2R$.
   c. $I = E/R$.
   d. $I = ER$.
   e. $R = EI$.
   f. $R = E/I$.

5. A 100-Ω resistor has 0.05 A flowing through it. How much voltage is dropped across this resistor?
   a. 0.5 mV.
   b. 5 mV.
   c. 2 V.
   d. 5 V.
   e. 200 V.

6. Which one of the following equations most directly determines the value of a resistance when you know the amount of voltage dropped across it and the amount of current flowing through it?
   a. $E = IR$.
   b. $E = I^2R$.
   c. $I = E/R$.
   d. $I = ER$.
   e. $R = EI$.
   f. $R = E/I$.

7. Which one of the following equations most directly determines the power dissipation of a resistance when you know the voltage drop and current?
   a. $P = IE$.
   b. $P = I^2R$.
   c. $P = E^2/R$.
   d. $I = P/E$.
   e. $E = P/I$.
   f. $R = E^2/P$.
   g. $R = P/I^2$.

8. Which one of the following equations most directly determines the power dissipation of a resistance when you know the voltage drop and amount of resistance?
   a. $P = IE$.
   b. $P = I^2R$.
   c. $P = E^2/R$.
   d. $I = P/E$.
   e. $E = P/I$.
   f. $R = E^2/P$.
   g. $R = P/I^2$.

3 THE SIMPLE ELECTRICAL CIRCUIT

9. Household light bulbs have a given power rating, and they operate at 120 V. Which one of the following equations most directly provides the current rating of these light bulbs?
   a. $P = IE$.
   b. $P = I^2R$.
   c. $P = E^2/R$.
   d. $I = P/E$.
   e. $E = P/I$.
   f. $R = E^2/P$.
   g. $R = P/I^2$.
   h. $I = E/R$.

10. How much current flows through a 60-W light bulb that is operating at 120 V?
    a. 0.5 A.
    b. 2 A.
    c. 180 A.
    d. 7200 A.

11. Which one of the following equations most directly leads to the resistance of an ordinary household light bulb, assuming that you know the power and voltage rating?
    a. $P = IE$.
    b. $P = I^2R$.
    c. $P = E^2/R$.
    d. $I = P/E$.
    e. $E = P/I$.
    f. $R = E^2/P$.
    g. $R = P/I^2$.

12. What is the power dissipation of a resistance that has 24 V dropped across it and 0.25 A flowing through it?
    a. 1.5 W.
    b. 6 W.
    c. 96 W.
    d. 144 W.

13. How much power is being dissipated by a 1000-Ω resistance that has 25 V applied to it?
    a. 0.025 W.
    b. 0.625 W.
    c. 40 W.
    d. 625 kW.

DC SERIES CIRCUITS

Dc Series Circuits

ABOUT THIS CHAPTER

This chapter contains a thorough description of the series circuit and its basic connections. It also explains how total resistance, total current, and voltage drops are determined in such a circuit. You will learn how to reduce a voltage to a desired level by the use of resistors and how to determine total resistance and current in a series circuit.

SERIES CIRCUITS

The amount of current flow is the same through every part of a series circuit.

A "series circuit" is an electrical circuit that provides only one path for current flow through all components in the circuit. Because there is just one path for current flow, it figures that the current must be the same through all parts of the circuit.

Figure 4-1 shows a series circuit that is composed of ten identical lamps (80 Ω apiece) and a 120-V source of electrical power. You will soon learn how to do the necessary calculations for determining the amount of current through the lamps. It is more important now to understand that the same amount of current—0.15 A—flows through all parts of the series circuit.

**Figure 4-1.
Series Lamp Circuit**

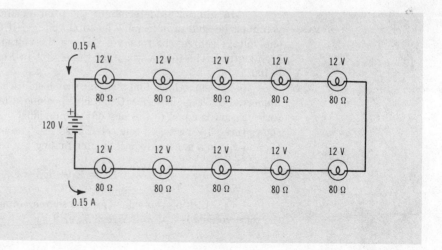

4 DC SERIES CIRCUITS

Total Resistance

The total resistance in a series circuit is equal to the sum of all the individual resistances. The mathematical symbol for total resistance in a circuit is R_T. *Figure 4-2* shows a series circuit that includes four resistances, labeled R_1, R_2, R_3, and R_4. The total resistance of the circuit can be indicated by this expression:

$$R_T = R_1 + R_2 + R_3 + R_4$$

Substituting the actual resistance values into that expression, you will find that the total resistance of the circuit is 100 Ω.

**Figure 4-2.
Total Resistance of Series Circuit**

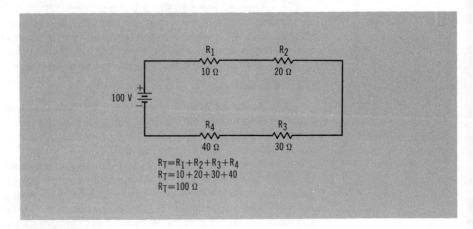

Total Voltage

In a series circuit that has only one voltage source, the total voltage is equal to the source voltage.

In addition to determining the total resistance of a series circuit, you must be able to determine its total voltage. If the circuit uses just one voltage source, the total voltage and the voltage rating of the source are identical. The total voltage for the circuit in *Figure 4-2*, for example, is 100 V.

You sometimes find a circuit that includes multiple voltage sources connected together in series. Generally speaking, the total voltage (E_T) in a series circuit is equal to the sum of the individual voltages. *Figure 4-3* shows three 3-V batteries connected in series. The batteries are labeled E_1, E_2, and E_3, so a general expression for finding the total voltage is:

$$E_T = E_1 + E_2 + E_3$$

Substituting the values for the source voltages, you can see that the total voltage is 3 V + 3 V + 3 V, or 9 V.

DC SERIES CIRCUITS

**Figure 4-3.
Batteries Connected in Series**

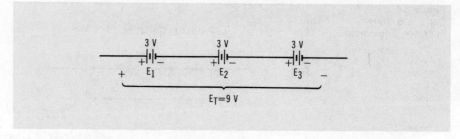

Series-Aiding Circuits

Two or more voltage sources connected so that they are series aiding produce a total voltage that is the sum of their individual voltages.

Notice, however, that the batteries in *Figure 4-3* are connected so that the negative terminal of one is connected to the positive terminal on the next. Connecting a series of two or more batteries in this fashion causes their individual voltages to aid one another. The polarity of the total voltage is the same as that of all the individual voltage sources. This is said to be a "series-aiding" circuit.

Series-Opposing Circuits

Two voltage sources connected so that they are series opposing produce a total voltage that is the difference between their individual voltages. The polarity of the total voltage is that of the larger voltage source.

Figure 4-4 shows two batteries connected in series, but with the negative terminals connected together. Used in this "series-opposing" circuit, the voltages tend to cancel. The total voltage is found by subtracting the individual voltages and assigning the overall polarity as that of the larger-valued voltage source. The difference between the voltages in this example is 6 V, and the polarity of the total voltage is the same as that of the 9-V battery.

**Figure 4-4.
Batteries in Series-Opposing Circuit**

Most flashlights use two 1.5-V dry cells to get a total lamp voltage of 3 V. That happens, though, only if you install the batteries so that their voltages aid one another. If you install one of the batteries backwards, the voltages will oppose one another and the total voltage will be zero.

A number of everyday portable electronic devices, such as portable radios, use four 1.5-V cells connected in series, giving a total voltage of 6 V. What is the total voltage if just one of them is connected in opposition to the others? The diagram in *Figure 4-5* demonstrates this situation.

4 DC SERIES CIRCUITS

**Figure 4-5.
Cells in Battery-operated Equipment**

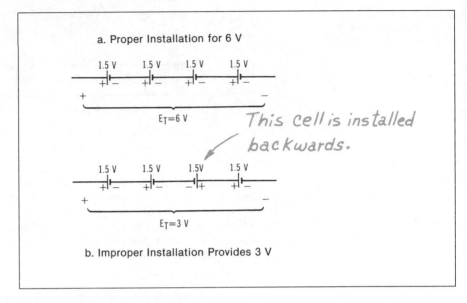

Total Current

You have already learned that the current in a series circuit is the same through all parts of the circuit. The question, though, is how much current flows through every part of a series circuit?

According to the basic form of Ohm's law, current is equal to voltage divided by resistance: $I = E/R$. You have just seen how you can determine the total voltage and total resistance in a series circuit. This information—with the help of Ohm's law—leads directly to the total amount of current:

$$I_T = \frac{E_T}{R_T}$$

The total current in a circuit can be found by dividing the total voltage by the total resistance.

This expression says that the total current flowing through a circuit is equal to the total voltage divided by the total resistance. So if you can determine total voltage and resistance in a series circuit, you can determine the amount of current flowing through the circuit.

Suppose you find that the total voltage in a circuit is 110 V and that the total resistance is 50 Ω. What is the total current?

$$I_T = \frac{E_T}{R_T}$$

$$I_T = \frac{110 \text{ V}}{50}$$

$$I_T = 2.2 \text{ A}$$

DC SERIES CIRCUITS

Voltage Drop

Recall that the voltage drop across a resistance is sometimes called an IR drop. That is, the voltage is determined by multiplying the amount of current flowing through the resistance by the value of the resistance. The task is quite simple if you already know the value of the resistance and have determined the total current through the circuit as described previously.

Amount of Voltage Drop

Figure 4-6 shows a series circuit composed of a 110-V source and two resistors, R_1 and R_2. What is the voltage drop across the two resistors?

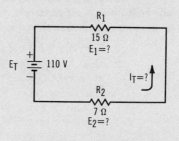

**Figure 4-6.
Voltage Drop Across Two Resistors**

In order to find the IR drop across the resistors, you must first determine the amount of current flowing through the circuit. And finding the total current means finding the total resistance.

To find the total resistance in this series circuit:

$$R_T = R_1 + R_2$$
$$R_T = 15 + 7$$
$$R_T = 22$$

To find the current through this circuit:

$$I_T = \frac{E_T}{R_T}$$
$$I_T = \frac{110 \text{ V}}{22}$$
$$I_T = 5 \text{ A}$$

Now that you know the current through the resistors, you are in a position to find the voltage dropped across each of them:

4 DC SERIES CIRCUITS

$$E_1 = I_T R_1$$
$$E_1 = 5 \text{ A} \times 15$$
$$E_1 = 65 \text{ V}$$

$$E_2 = I_T R_2$$
$$E_2 = 5 \text{ A} \times 7$$
$$E_2 = 35 \text{ V}$$

The sum of all IR voltage drops in a series circuit is equal to the total voltage applied to the circuit.

The voltage drops across resistors R_1 and R_2 are 65 V and 35 V, respectively. Notice that the voltage drops in this series circuit add up to the total source voltage: 65 V + 35 V = 110 V. The sum of all voltage drops in a series circuit is equal to the total source voltage.

Assuming that you know the total source voltage and the values of all resistors in a series circuit, you can use this general procedure to determine the amount of voltage across each resistor:

1. Sum the values of all the resistors to obtain the total resistance, R_T.
2. Use the equation $I_T = E_T/R_T$ to determine the total current (amount of current flowing through each resistor in a series circuit).
3. Use the equation $E = IR$ to calculate the amount of voltage dropped across each resistor, where E is the voltage dropped across a resistor having resistance R and current I flowing through it.

You can double-check your result by summing the IR drops across each resistor. The sum should be equal to the source voltage. If that is not the case, you have made an error somewhere along the way.

Polarity of Voltage Drop

It is often just as important to determine the polarity of a voltage dropped across a resistor as it is to find the amount of voltage. To do this, recall that electron current leaves the negative terminal of a voltage source, flows through the circuit, and returns to the positive terminal. The direction of current flow through a series circuit determines the polarity of the IR voltage drops across each resistance in the circuit.

The point where electron current enters a resistance is negative compared to the point where current leaves the resistance.

Voltage always exists between two different points—never at one point only. Therefore, voltage is expressed as being across two points in a circuit or in terms of one point with respect to another point. The point at which electron current enters a resistance is negative with respect to the point at which it leaves. *Figure 4-7* illustrates this important rule.

DC SERIES CIRCUITS

**Figure 4-7.
Polarity of Voltage
Drops**

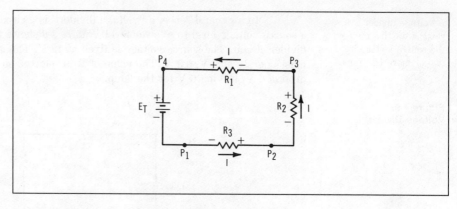

Voltage Reference Point (Ground)

A ground point shown on a schematic is a point of reference for measuring voltages in the circuit.

A voltage ground is often used to establish a reference point for all other voltage readings. A ground point is the reference point for measuring all voltages in a circuit. Voltages listed on a schematic diagram are read with respect to the ground point.

The ground, or voltage reference point, for a circuit is indicated by a symbol that resembles an inverted pyramid. There are two ground symbols shown in *Figure 4-8*. Electron current flows from P_1 to P_2 to P_3 through the voltage source to ground, and from ground to P_1. In this case, I_T is equal to 1 A. This means P_1 is $+16$ V with respect to ground. P_2 is $+14$ V with respect to P_1 (and thus $+30$ V with respect to ground). P_3 is $+50$ V with respect to ground.

**Figure 4-8.
Ground Is Reference
Point**

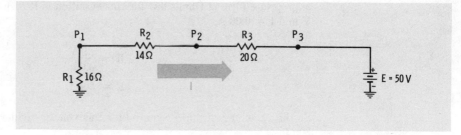

Voltage Dividers

You have seen that current is the same through every component in a series circuit. You have also learned that the source voltage is divided among the resistances in a circuit. The fact that a source voltage can be divided among resistances in a series circuit gives rise to the notion of voltage dividers.

4 DC SERIES CIRCUITS

A voltage divider is a series circuit that divides the source voltage into two or more IR drops.

In a general sense, a "voltage divider" is a circuit that steps down a source voltage level to a lower level. *Figure 4-9* shows a simple voltage divider circuit. The source voltage is fixed at 12 V. The lamp, however, must operate at 6 V, 0.3 A. The value of R is selected so that it has an IR drop of 6 V, leaving 6 V for the lamp.

**Figure 4-9.
Voltage Divider**

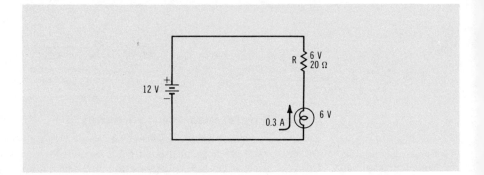

To get a practical appreciation of voltage divider circuits, suppose that you want to operates 3-V portable radio from the 12-V system in your automobile. As shown in Figure 4-10a, you must first measure the amount of current required for operating the radio from its normal 3-V source. The diagram indicates that the radio uses 30 mA, or 0.03 A.

What value of resistor must be inserted in series with the radio in order to make it operate properly from a 12-V source? Or to put it another way, what value resistance is necessary for dropping 9 V at 30 mA? The appropriate form of Ohm's law for this situation is R = E/I, where E = 9 V and I = 0.03 A.

$$R = \frac{9 \text{ V}}{0.03 \text{ A}}$$

$$R = 300 \text{ }\Omega$$

So the value of resistance required for this voltage-divider operation is 300 Ω.

Recall that resistors are also specified according to their power ratings. What is the minimum power rating of the resistor in this situation? A useful power formula for this situation is P = IE, where I = 0.03 A and E = 9 V.

$$P = 0.03 \text{ A} \times 9 \text{ V}$$

$$P = 0.27 \text{ W}$$

Fixed resistors are available in ¼-W, ½-W, 1-W, and 2-W ratings. The dropping resistor will dissipate 0.27 W, or a bit more than ¼ W. So you should use one having at least a ½-W rating.

DC SERIES CIRCUITS

Figure 4-10.
Using a Voltage Divider

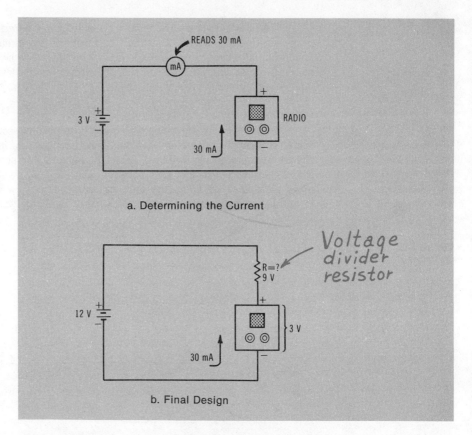

a. Determining the Current

b. Final Design

Now suppose that you have completed this bit of electrical engineering based on your understanding of series voltage-divider circuits and Ohm's law. You know from the calcuations that you need a 300-Ω, ½-W resistor. But you cannot find a resistor having those specifications. (That happens to be a non-standard set of values.) The best you can do is a 270-Ω or a 330-Ω resistor.

You have to make an important, but common, engineering decision. If you use a 270-Ω resistor, the voltage to the radio will be slightly greater than 3 V. On the other hand, if you use a 330-Ω resistor, the voltage to the radio will be slightly less. In a practical sense, either value will work—the error is no more than 10% in either case.

Some kinds of circuits that require a voltage divider cannot tolerate an error of 10%. When that is the case, you can use a variable resistor for the dropping resistor. If you need exactly 300-Ω resistance, locate a 500-Ω variable resistor (that is a standard value for variable resistors, or potentiometers). Before connecting it into the circuit, use an ohmmeter to set the resistance to exactly 300 Ω.

4 DC SERIES CIRCUITS

Voltage dividers are used in other ways for a variety of different purposes. One of the skills that must be mastered by an electronics technician is to recognize a voltage-divider circuit when he or she sees one on a schematic diagram. Recognizing the circuit readily leads to an understanding of how it should behave.

HOW SWITCHES ARE USED IN SERIES CIRCUITS

Switches can be connected in series to perform some useful tasks and to explain the operation of certain logic circuits used in computers. *Figure 4-11* shows two switches, A and B, connected in series. *Figure 4-11a* shows that current cannot flow between points P_1 and P_2 as long as either or both switches are open. The only way current can flow between points P_1 and P_2 is when both switches are closed at the same time, as shown in *Figure 4-11b*.

> Current can flow through switches connected in series only when all contacts are closed.

**Figure 4-11.
Switches in Series**

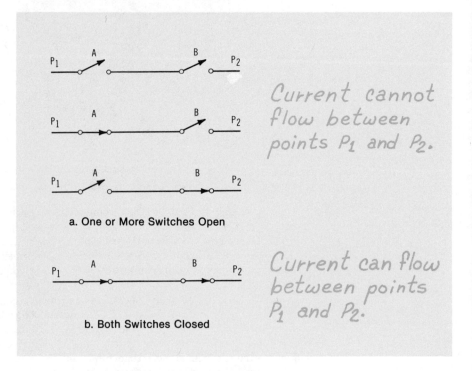

A switch circuit that works according to this principle is called an AND circuit. It has that name because current flows only when switches A *AND* B are closed at the same time.

WHAT HAVE WE LEARNED?

1. A series circuit provides a single path for current flow through all of its components.
2. The same current flows through all parts of a series circuit.

DC SERIES CIRCUITS 4

3. The total resistance of a series circuit is equal to the sum of the individual resistances.
4. Series-aiding voltage sources increase the total voltage; series-opposing voltage sources decrease the total voltage.
5. The total current in a circuit can be found by Ohm's law: divide the total voltage by the total resistance.
6. The voltage across a resistor can be found by Ohm's law: multiply the resistance value times the amount of current flowing through the resistor. This voltage drop is sometimes called an IR drop.
7. The sum of all IR voltage drops in a series circuit is equal to the total voltage.
8. The point where electron current enters a resistance is negative compared to the point where the current leaves that resistance.
9. Voltage is always measured with respect to two given points in a circuit.
10. Voltage ground is often taken as the reference point for measuring all voltages in a circuit.
11. A voltage divider is a series circuit that divides a source voltage into two or more smaller voltage levels.
12. Where two or more switches are connected in series, current can flow through the circuit only when all switch contacts are closed. This is sometimes called an AND circuit.

KEY WORDS

AND logic
Ground reference
IR voltage drop
Series aiding

Series circuit
Series opposing
Voltage divider

4 DC SERIES CIRCUITS

Quiz for Chapter 4

1. Which one of the following statements is true for current flowing in a series circuit?
   a. The amount of current flow through each part of a series circuit can be different, depending on the resistance of each part and the amount of voltage applied to it.
   b. The total current in a series circuit is equal to the sum of the currents flowing through the individual components in the circuit.
   c. The same current flows through every part of a series circuit.
   d. The total current in a series circuit is equal to the total voltage multiplied by the total resistance.

2. Which one of the following statements is true for resistance in a series circuit?
   a. The total resistance is equal to the sum of the individual resistances in a series circuit.
   b. The total resistance of a series circuit is always less than the value of the smallest resistance.
   c. The total resistance in a series circuit is equal to the average value of the individual resistances.
   d. The total resistance of a series circuit is equal to the total current divided by the total voltage.

3. What is the total resistance of a series circuit that contains four resistors having values of 220 Ω, 470 Ω, 1 Ω, and 1.2 Ω?
   a. 692.2 Ω.
   b. 722.5 Ω.
   c. 2890 Ω.
   d. 4870 Ω.

4. Which one of the following statements is true for voltage in a series circuit?
   a. The total voltage is equal to the sum of the individual voltages in a series circuit.
   b. The total voltage in a series circuit is always less than the value of the smallest voltage.
   c. The total voltage in a series circuit is equal to the average value of the individual voltages.
   d. The total voltage of a series circuit is equal to the total current multiplied by the total resistance.

5. A circuit has two different voltage sources that are connected in a series-aiding form. If the sources are rated at 6 V and 9 V, what is the total source voltage?
   a. 3 V.
   b. 6 V.
   c. 7.5 V.
   d. 9 V.
   e. 15 V.

6. A circuit has two different voltage sources that are connected in a series-opposing form. If the sources are rated at 6 V and 9 V, what is the total source voltage?
   a. 3 V.
   b. 6 V.
   c. 7.5 V.
   d. 9 V.
   e. 15 V.

DC SERIES CIRCUITS 4

**Figure 4-12.
Series Circuit with
Three Resistors**

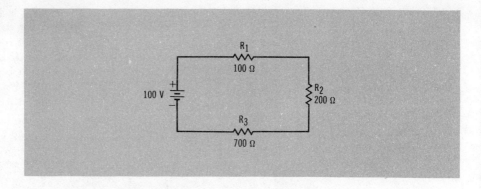

7. What is the total resistance of the circuit in *Figure 4-12*?
   a. 86 Ω.
   b. 333 Ω.
   c. 700 Ω.
   d. 1000 Ω. ✓

8. What is the total current through the circuit in *Figure 4-12*?
   a. 0.1 A. ✓
   b. 0.3 A.
   c. 1.16 A.
   d. 100 A.

9. What is the voltage drop across resistor R_1 in *Figure 4-12*?
   a. 1 V.
   b. 10 V.
   c. 11.6 V.
   d. 30 V.

e. 33.3 V.
f. 100 V.

10. What is the IR drop across resistor R_2 in *Figure 4-12*?
    a. 2 V.
    b. 5 V.
    c. 20 V.
    d. 24.2 V.
    e. 33.3 V.
    f. 100 V.

11. What is the voltage drop across R_3 in *Figure 4-12*?
    a. 0.33 V.
    b. 21 V.
    c. 33.3 V.
    d. 70 V.
    e. 100 V.

**Figure 4-13.
Circuit for Quiz
Questions 12 and 13**

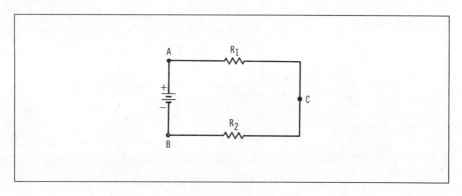

UNDERSTANDING ELECTRICITY AND ELECTRONICS CIRCUITS

4 DC SERIES CIRCUITS

12. Point A in *Figure 4-13* is:
  a. positive with respect to point B.
  b. negative with respect to point B.
  c. negative with respect to point C.

13. Point C in *Figure 4-13* is:
  a. positive with respect to point B.
  b. negative with respect to point B.
  c. positive with respect to point A.

Figure 4-14.
Circuit for Quiz Questions 14–16

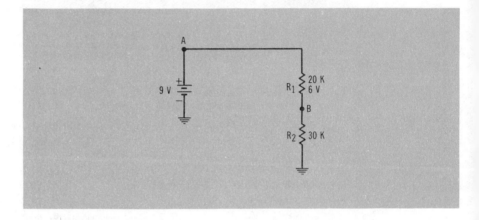

14. What is the IR drop across resistor R_2 in *Figure 4-14*?
  a. 3 V.
  b. 6 V.
  c. 9 V.
  d. 15 V.

15. What is the voltage at Point A with respect to ground in *Figure 4-14*?
  a. 0 V.
  b. 3 V.
  c. 4.5 V.
  d. 6 V.
  e. 9 V.
  f. 15 V.

16. What is the voltage at Point B with respect to ground in *Figure 4-14*?
  a. 0 V.
  b. 3 V.
  c. 4.5 V.
  d. 6 V.
  e. 9 V.
  f. 15 V.

Dc Parallel Circuits

ABOUT THIS CHAPTER

This chapter explains dc parallel circuits. You will learn to recognize parallel circuits, determine currents, voltages, and total resistance, and we will explore applications of such circuits. You will also be introduced to the notion of developing equivalent circuits.

PARALLEL CIRCUITS

A "parallel circuit" is an electrical circuit that contains two or more separate paths for current flow. The separate paths for current flow in a parallel circuit are called "branches." The parallel circuit illustrated in *Figure 5-1* has three separate branches. Also notice that the two ends of each branch are connected to the two ends of the other branches.

A parallel circuit is an electrical circuit that has two or more separate paths, or branches, for current flow.

Figure 5-1. Parallel Circuit

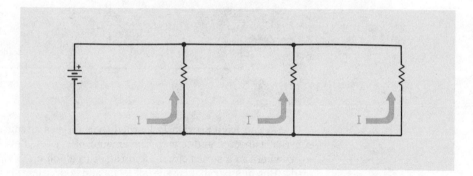

In a parallel circuit, the total current is divided among the individual branches so the voltage is the same across each branch.

Figure 5-2 shows another example of a parallel circuit. The circuit is a string of Christmas tree lamps that permits one or more lamps in the circuit to be removed (or burned out), while the others continue to operate. The reason for this effect is that each lamp has its own path for current flow; interrupting the path for current flow to any of the lamps does not affect the others. Each lamp has the same voltage applied to it because the total current is divided evenly among branches.

Another form of parallel circuit uses more than one voltage source in parallel in order to increase the availability of current. The arrangement of sources shown in *Figure 5-3* is intended for this purpose. Also, the greater the amount of battery energy available, the longer the source can operate at a given current and voltage level.

5 DC PARALLEL CIRCUITS

**Figure 5-2.
Lamps in Parallel**

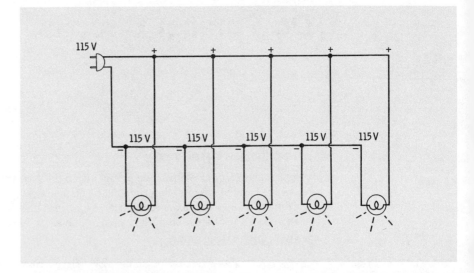

**Figure 5-3.
Batteries in Parallel**

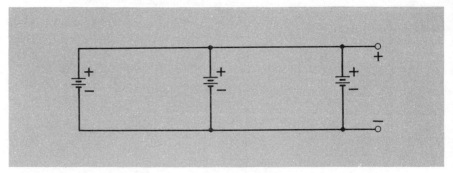

You have previously learned that a series circuit has all loads and sources connected end-to-end. The amount of current flow is the same everywhere in a series circuit, and the source voltage divides among the loads. But in a parallel circuit, all loads and sources are connected across the same two points, the total current divides among the loads, and the voltage is the same for every source and load. Compare the series and parallel circuits in *Figure 5-4*.

Current Flow

Current in each branch of a parallel circuit must originate from the same source. This means that different groups of electrons flow through each branch. The currents through different branches in a parallel circuit will be equal only if the resistance is the same in each branch. The source must supply current to all branches in a parallel circuit, so the total current is the sum of all the branch currents, as is shown in *Figure 5-5*.

The total current (I_T) in a parallel circuit is equal to the sum of the currents flowing through all branches of the circuit.

DC PARALLEL CIRCUITS

Figure 5-4.
Two Basic Circuits

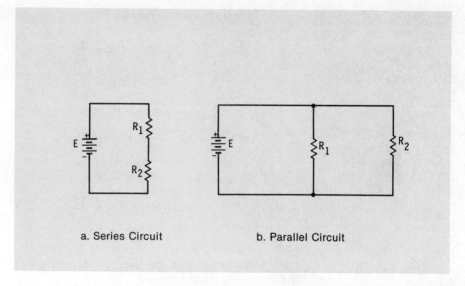

a. Series Circuit   b. Parallel Circuit

Figure 5-5.
Total Current in Parallel Circuit

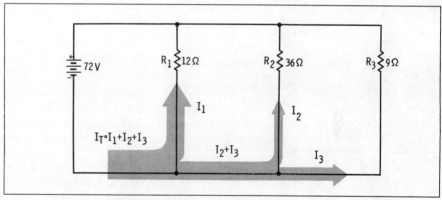

Finding the amount of current flowing through the individual branches of a parallel circuit is a matter of applying Ohm's law, $I = E/R$. Given the values for source voltage and resistance in *Figure 5-5*, the current through resistor R_1 is equal to the source voltage (72 V) divided by the resistance value of that resistor (12 Ω):

$$I_1 = \frac{E}{R_1}$$

$$I_1 = \frac{72 \text{ V}}{12}$$

$$I_1 = 6 \text{ A}$$

5 DC PARALLEL CIRCUITS

To find the current through R_2:

$$I_2 = \frac{E}{R_2}$$

$$I_2 = \frac{72\text{ V}}{36}$$

$$I_2 = 2\text{ A}$$

And to find the current through R_3:

$$I_3 = \frac{E}{R_3}$$

$$I_3 = \frac{72\text{ V}}{9}$$

$$I_3 = 8\text{ A}$$

The total current can then be determined by summing the current through the individual branches:

$$I_T = I_1 + I_2 + I_3$$
$$I_T = 6\text{ A} + 2\text{ A} + 8\text{ A}$$
$$I_T = 16\text{ A}$$

Total Resistance

The total resistance of a parallel circuit is always less than the resistance in any branch of the circuit.

In a series circuit, total resistance is found by summing the values of the individual resistances. That is not the case for parallel circuits, where the total resistance is always less than the value of resistance in any branch.

There are two entirely different approaches to finding the total resistance of a series circuit. The first is based on the fact that the total resistance is equal to the total voltage divided by the total current (a form of Ohm's law). The other approach deals only with the resistances themselves.

Assuming that you know (or can determine) the values of total voltage (E_T) and total current (I_T) in a parallel circuit, you can determine the total resistance (R_T) by Ohm's law: $R_T = E_T/I_T$.

Figure 5-6 illustrates the first approach. This approach determines total resistance according to this form of Ohm's law: $R_T = E_T/I_T$. To use that equation assumes that you know the total current flow, I_T. You have previously learned how to determine the total current flow in a parallel circuit. The procedure here is to determine the individual currents, sum those currents to find the total current, and divide the total current into the total voltage.

DC PARALLEL CIRCUITS

**Figure 5-6.
Using Ohm's Law to
Determine Total
Resistance**

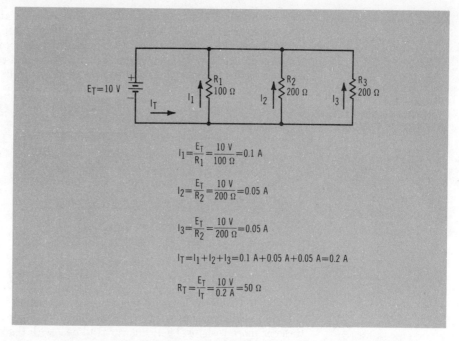

$$I_1 = \frac{E_T}{R_1} = \frac{10\ V}{100\ \Omega} = 0.1\ A$$

$$I_2 = \frac{E_T}{R_2} = \frac{10\ V}{200\ \Omega} = 0.05\ A$$

$$I_3 = \frac{E_T}{R_2} = \frac{10\ V}{200\ \Omega} = 0.05\ A$$

$$I_T = I_1 + I_2 + I_3 = 0.1\ A + 0.05\ A + 0.05\ A = 0.2\ A$$

$$R_T = \frac{E_T}{I_T} = \frac{10\ V}{0.2\ A} = 50\ \Omega$$

The fundamental equation for determining the total resistance of a parallel circuit is rather difficult to use, even with the help of an electronic calculator. The equation looks like this:

$$R_T = \frac{1}{[(1/R_1) + (1/R_2) + (1/R_3)\ldots]}$$

where R_T is the total resistance and R_1, R_2, R_3, and so on, are the values of the individual resistors. The general procedure is to divide each resistance value into 1, sum the results, then divide that sum into 1.

Product-Over-Sum Rule

For two resistances in parallel, the product-over-sum rule says that their combined resistance is equal to the product of their two values divided by the sum of the two values.

An alternative approach for finding the total resistance of a parallel circuit is based on the fact that the total resistance of two resistors is equal to the product of the two values divided by their sum. The mathematical version of that statement is:

$$R_T = \frac{(R_1 \times R_2)}{(R_1 + R_2)}$$

where R_T is the combined resistance of two resistors, R_1 and R_2, that are connected in parallel. This is often called the "product-over-sum rule."

5 DC PARALLEL CIRCUITS

Figure 5-7 suggests an example. The values of the two parallel-connected resistors are 40 Ω and 60 Ω. Applying the product-over-sum rule for total resistance, you find that the combined resistance is 24 Ω. Notice that you do not have to know the values of voltage and current in the circuit in order to determine total resistance in this fashion.

**Figure 5-7.
Total Resistance of Two Resistors in Parallel**

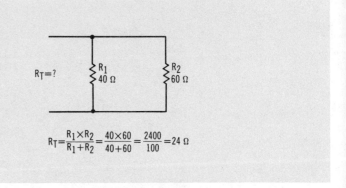

But what if there are more than two resistors in a parallel circuit? How can you use the product-over-sum rule when a parallel circuit includes three or more resistors? The answer is to deal with the resistors two at a time. *Figure 5-8* illustrates the procedure for a circuit having three resistors in parallel.

Figure 5-8a shows the original circuit composed of resistors having values of 60 Ω, 40 Ω, and 120 Ω. The first step in the procedure is to select two of the resistors and apply the product-over-sum rule to them in order to determine an equivalent resistance for the two. In this case, the 40-Ω and 120-Ω resistors combine to yield an equivalent resistance, R_A, of 30 Ω.

The step in *Figure 5-8b* replaces those two resistors with one of their equivalent value. Now there are just two resistors remaining in the parallel circuit. Applying the product-over-sum rule to those resistors determines the total resistance of the circuit, 20 Ω.

Using equivalent resistances makes it possible to apply the product-over-sum rule to determine the total resistance of a circuit having any number of resistors connected in parallel.

**Figure 5-8.
Total Resistance of
Three Resistors in
Parallel**

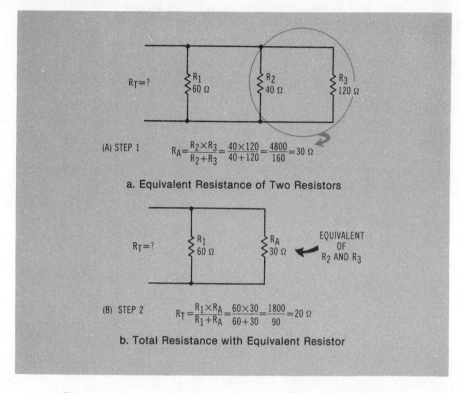

(A) STEP 1 $\quad R_A = \dfrac{R_2 \times R_3}{R_2 + R_3} = \dfrac{40 \times 120}{40 + 120} = \dfrac{4800}{160} = 30\ \Omega$

a. Equivalent Resistance of Two Resistors

(B) STEP 2 $\quad R_T = \dfrac{R_1 \times R_A}{R_1 + R_A} = \dfrac{60 \times 30}{60 + 30} = \dfrac{1800}{90} = 20\ \Omega$

b. Total Resistance with Equivalent Resistor

Figure 5-9 illustrates the equivalent-resistance procedure for a parallel circuit comprising four different resistors. The first step combines resistors R_3 and R_4 to get an equivalent resistance, R_A. The second step combines resistor R_2 with equivalent resistance R_A to produce another equivalent resistance, R_B. The final step combines resistor R_1 with equivalent resistance R_B to get the total resistance of the circuit.

**Figure 5-9.
Using Equivalent
Resistances to
Determine Total
Resistance**

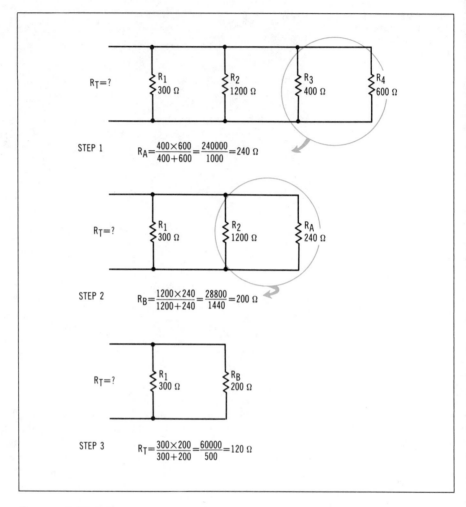

Current Dividers

A current divider is a parallel circuit that divides the source current into two or more levels of current for a specific application.

You learned in Chapter 4 that resistors connected in series make up a voltage divider. Now we will see that resistors connected in parallel make up a "current divider." *Figure 5-10* shows examples of voltage and current dividers.

The circuit in *Figure 5-11* shows a pair of resistors connected in parallel. As a current divider, one branch carries 40 mA of current while the other carries 5 mA. Of course, the total current is 45 mA. If the value of the resistor on the 40-mA branch is 220 Ω, what value resistance should you use to get the desired 5 mA to flow in the other branch? One way to go about determining the value of R_2 is to find the voltage across it, and then divide that voltage by the current through it.

DC PARALLEL CIRCUITS

**Figure 5-10.
Resistors in Divider
Circuits**

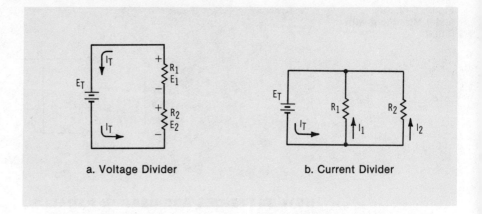

a. Voltage Divider   b. Current Divider

**Figure 5-11.
Determining Resistor
Value in Current-Divider
Circuit**

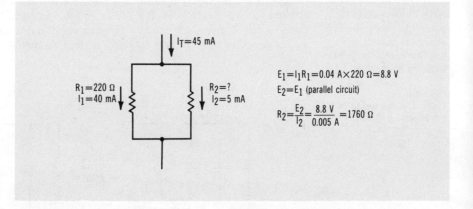

Meter movements use current dividers. A meter movement typically shows full-scale deflection when a current of 10 μA flows through it. Forcing more than full-scale current through a movement will destroy it. So how is it possible to use such a meter movement to measure larger amounts of current—1 A, for instance? The solution, as illustrated in *Figure 5-12*, is to divide the higher current so that 10 μA goes through the meter movement and the remainder (1 A minus 10 μA) is bypassed through another resistor. A bypass resistor that is used for this purpose is called a "shunt" resistor.

**Figure 5-12.
Meter Movement with Shunt**

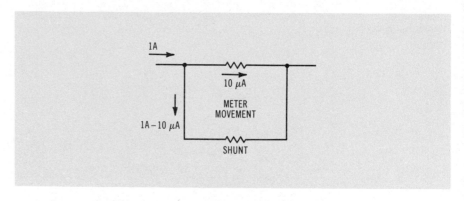

HOW BATTERIES ARE USED IN PARALLEL

Connecting cells or batteries of like voltage in parallel increases the amount of current and lengthens the time before they are discharged.

For heavy-duty operation, batteries of equal voltage are constructed with their individual cells connected in parallel. This enables them to deliver more current at a given moment and to deliver power for a longer period of time. *Figure 5-13* shows some examples.

**Figure 5-13.
Cells and Batteries in Parallel**

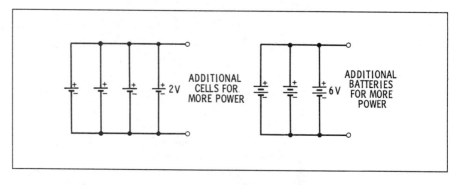

HOW SWITCHES ARE USED IN PARALLEL

Connecting switches in parallel makes it possible to complete a path for current flow when only one switch is closed.

Recall that all switches connected in a series circuit must be closed before current can flow through the circuit. *Figure 5-14* shows how such a circuit works when the switches are connected in parallel. As long as both switches are open, current cannot flow between points P1 and P2. However, closing either switch (or both), allows current to flow.

Switches connected in parallel form an OR circuit, so-called because it allows current to flow when one *OR* the other *OR* both are closed.

DC PARALLEL CIRCUITS

**Figure 5-14.
Switches in Parallel**

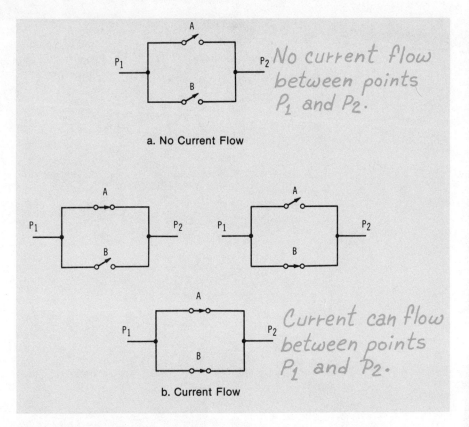

a. No Current Flow

No current flow between points P_1 and P_2.

b. Current Flow

Current can flow between points P_1 and P_2.

WHAT HAVE WE LEARNED?

1. A parallel circuit has two or more separate paths for current flow.
2. In a parallel circuit, the total current is divided among the individual branches; the voltage is the same across each branch.
3. The total current in a parallel circuit is equal to the sum of the currents in the individual branches.
4. The total resistance of a parallel circuit is always less than the value of the smallest resistance.
5. You can determine the total resistance of a parallel circuit by Ohm's law: the total resistance is equal to the voltage divided by the total current.
6. A current divider is a parallel circuit that divides a source current into two lesser amounts of current.
7. Connecting cells or batteries in parallel extends their useful life or makes more current available to the circuit.
8. Where switches are connected in parallel, closing just one, or any combination of them, allows current to flow through the circuit. This circuit is sometimes called an OR circuit.

5 DC PARALLEL CIRCUITS

KEY WORDS

Current divider
Equivalent resistance
OR circuit

Parallel circuit
Product-over-sum rule
Shunt resistor

DC PARALLEL CIRCUITS

Quiz for Chapter 5

1. Which one of the following statements is true for current flowing in a parallel circuit?
   a. The amount of current flow through each branch of a parallel circuit can be different, depending on the resistance of each branch part and the amount of voltage applied to it.
   b. The total current in a parallel circuit is always less than the smallest amount of current.
   c. The same current always flows through every part of a parallel circuit.
   d. The total current in a parallel circuit is equal to the total voltage multiplied by the total resistance.

2. Which one of the following statements is true for resistance in a parallel circuit?
   a. The total resistance is equal to the sum of the individual resistances in a parallel circuit.
   b. The total resistance of a parallel circuit is always less than the value of the smallest resistance.
   c. The total resistance in a parallel circuit is equal to the average value of the individual resistances.
   d. The total resistance of a parallel circuit is equal to the total current divided by the total voltage.

3. What is the total resistance of a parallel circuit that contains two resistors having values of 220 Ω and 470 Ω?
   a. 110 Ω.
   b. 150 Ω.
   c. 345 Ω.
   d. 690 Ω.

4. Which one of the following statements is true for voltage in a parallel circuit?
   a. The total voltage is equal to the sum of the voltages across the individual branches in a parallel circuit.
   b. The total voltage in a parallel circuit is always less than the value of the smallest voltage.
   c. The total voltage in a parallel circuit is equal to the average value of the individual voltages.
   d. The total voltage of a parallel circuit is the same as the voltages across each branch.

5. What is the voltage drop across resistor R_1 in *Figure 5-15*?
   a. 8 V.
   b. 12 V.
   c. 16 V.
   d. 20 V.
   e. 25 V.

6. What is the voltage drop across resistor R_2 in *Figure 5-15*?
   a. 8 V.
   b. 12 V.
   c. 16 V.
   d. 20 V.
   e. 25 V.

**Figure 5-15.
Circuit for Quiz
Questions 5-10.**

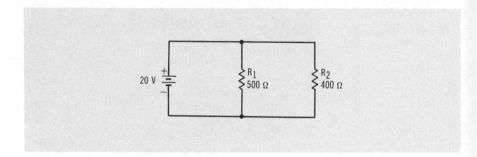

7. What is the current through resistor R_1 in *Figure 5-15*?
   a. 22 mA.
   b. 40 mA.
   c. 50 mA.
   d. 90 mA.
   e. 1.2 A.

8. What is the current through resistor R_2 in *Figure 5-15*?
   a. 22 mA.
   b. 40 mA.
   c. 50 mA.
   d. 90 mA.
   e. 1.2 A.

9. What is the total circuit current, R_T, in *Figure 5-15*?
   a. 22 mA.
   b. 40 mA.
   c. 50 mA.
   d. 90 mA.
   e. 1.2 A.

10. What is the total resistance of the circuit in *Figure 5-15*?
    a. 22.2 Ω.
    b. 222 Ω.
    c. 400 Ω.
    d. 450 Ω.
    e. 500 Ω.
    f. 900 Ω.

11. What is the voltage across the 1000-Ω resistor in *Figure 5-16*?
    a. 20 V.
    b. 40 V.
    c. 60 V.
    d. 80 V.
    e. 120 V.

12. What is the voltage across the 2000-Ω resistor in *Figure 5-16*?
    a. 20 V.
    b. 40 V.
    c. 60 V.
    d. 80 V.
    e. 120 V.

**Figure 5-16.
Circuit for Quiz
Questions 11-18**

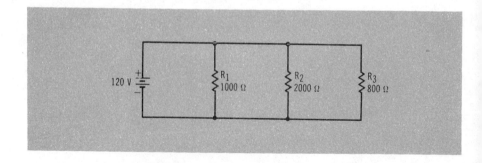

DC PARALLEL CIRCUITS

13. What is the voltage across the 800-Ω resistor in *Figure 5-16*?
    a. 20 V.
    b. 40 V.
    c. 60 V.
    d. 80 V.
    e. 120 V.

14. What is the current through resistor R_1 in *Figure 5-16*?
    a. 31.6 mA.
    b. 40 mA.
    c. 60 mA.
    d. 120 mA.
    e. 150 mA.
    f. 330 mA.

15. What is the current through resistor R_2 in Figure 5-16?
    a. 31.6 mA.
    b. 40 mA.
    c. 60 mA.
    d. 120 mA.
    e. 150 mA.
    f. 330 mA.

16. What is the current through resistor R_3 in *Figure 5-16*?
    a. 31.6 mA.
    b. 40 mA.
    c. 60 mA.
    d. 120 mA.
    e. 150 mA.
    f. 330 mA.

17. What is the total current for the circuit in *Figure 5-16*?
    a. 31.6 mA.
    b. 40 mA.
    c. 60 mA.
    d. 120 mA.
    e. 150 mA.
    f. 330 mA.

18. What is the total resistance of the circuit in *Figure 5-16*?
    a. 120 Ω.
    b. 364 Ω.
    c. 622 Ω.
    d. 1200 Ω.
    e. 1400 Ω.
    f. 3800 Ω.

Combined Series and Parallel Circuits

ABOUT THIS CHAPTER

In Chapters 4 and 5 we analyzed dc circuits on the basis of their voltages, currents, and resistances. Those discussions, however, separated the principles in terms of series circuits and parallel circuits. The purpose of this chapter is to show you how to analyze circuits that contain both series and parallel paths for current flow.

You will first learn to separate the series and parallel portions of a circuit and then to analyze those portions individually.

A good understanding of the material in this chapter will prepare you for dealing with complex electronic circuits and applying the step-by-step thinking required for working with such circuits.

COMBINATION CIRCUITS

Figure 6-1 shows the distribution of current in basic series and parallel circuits. The same current flows through both resistances in the series circuit, whereas the current divides between the two resistances in the parallel circuit. You have already learned how to analyze the distribution of current among the components in such circuits.

A combination circuit is an electrical circuit that includes both series and parallel paths for current flow. Consider the examples in *Figure 6-2*. In the first circuit, resistors R_2 and R_3 provide separate paths for current flow, so you can conclude that those two resistances are connected in parallel with one another. The combined currents for resistors R_2 and R_3 flow through resistor R_1. So you know R_1 is connected in series with the other two.

6 COMBINED SERIES AND PARALLEL CIRCUITS

**Figure 6-1.
Currents in Series and
Parallel Circuits**

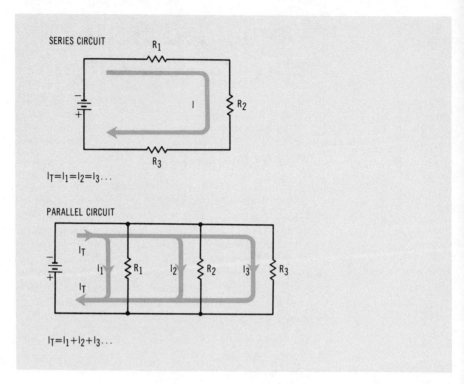

**Figure 6-2.
Currents in
Combination Circuits**

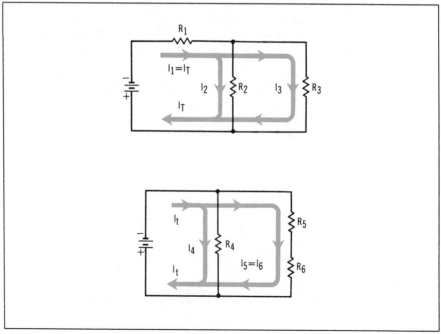

COMBINED SERIES AND PARALLEL CIRCUITS

The second circuit in *Figure 6-2* is also a combination circuit. It is basically a parallel circuit composed of two separate branches—part of the current from the source flows through resistor R_4 and the remainder flows through the series combination of R_5 and R_6. Resistors R_5 and R_6 are connected in series with one another; the same current flows through them.

There are many different forms of combination circuits. Before you can properly analyze the distribution of voltages and resistance values, you must distinguish series-connected elements from elements connected in parallel. Once you have made those distinctions, you can apply the basic principles of series and parallel circuits.

Total Resistance

You have already seen how you can analyze a parallel circuit with the help of equivalent resistances—a resistance made up of two or more other resistances. It is possible to reduce any resistive circuit to a single equivalent resistance. Equivalent resistances play a vital role in the analysis of combination circuits.

Figure 6-3 shows a combination circuit composed of three resistors. Resistor R_1 is connected in series with a parallel arrangement of resistors R_2 and R_3. You can apply the product-over-sum rule to resistors R_2 and R_3 to determine the value of an equivalent resistance, R_A. This equivalent resistance then appears in series with R_1, so the total resistance of the circuit is equal to the value of R_1 plus the value of R_A.

Figure 6-3.
Total Resistance, Simple Combination Circuit

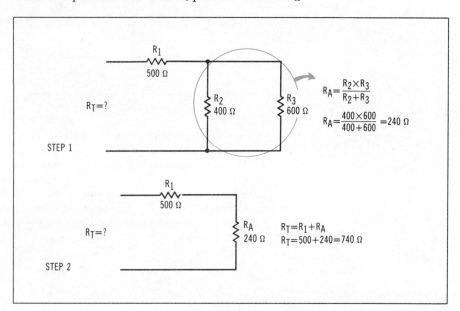

6 COMBINED SERIES AND PARALLEL CIRCUITS

Figure 6-4 shows another combination circuit that is composed of three resistors. However, the arrangement of these three resistors is quite different from the previous example. Resistors R_1 and R_2 are connected in series with one another. Therefore, you can combine them into an equivalent resistance, R_A, by summing their values. Substituting R_A for resistors R_1 and R_2, we are left with a simple parallel circuit composed of R_A and resistor R_3. The product-over-sum rule for those two resistances yields the total resistance of the circuit.

Figure 6-4.
Total Resistance, Simple Combination Circuit

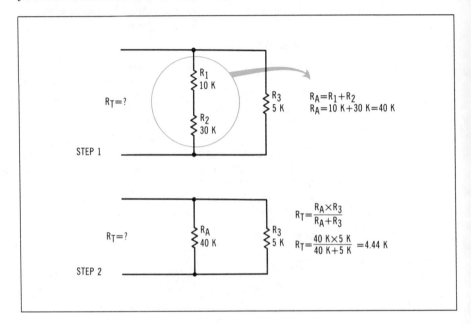

Analyzing Resistance

Where do you begin this kind of analysis of a combination circuit? Wherever you can. Find a pair of resistors connected in parallel and determine their equivalent resistance. Or find some resistors connected in series and sum their values to get an equivalent resistance. Conduct the procedure one step at a time, and you will eventually end up with a single equivalent resistance that represents the total resistance of the circuit. *Figure 6-5* illustrates this procedure for a more complicated combination circuit.

COMBINED SERIES AND PARALLEL CIRCUITS

**Figure 6-5.
Total Resistance,
Complicated
Combination Circuit**

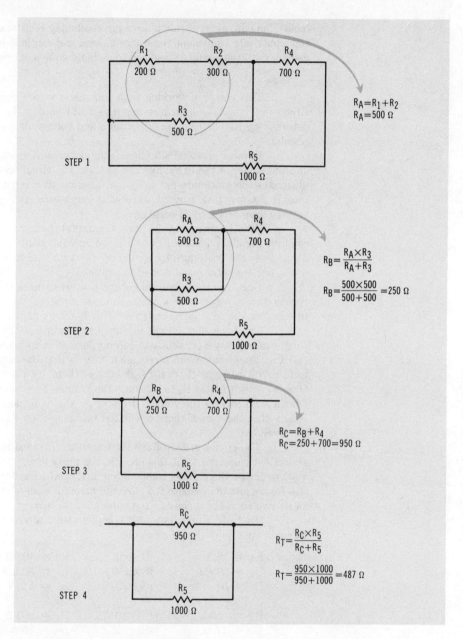

Total Current and Voltage

You have already seen that there is no fixed procedure for determining the total resistance of combination circuits. The exact procedure depends on the arrangement of the components in the circuit at

6 COMBINED SERIES AND PARALLEL CIRCUITS

hand. The only general rule is to begin combining resistances wherever you can, substitute equivalent resistance values, and continue in a systematic fashion until the circuit is reduced to a single equivalent resistance.

Analyzing the Circuit

This notion of working with equivalent values and approaching the circuit in a step-by-step fashion applies equally well to the procedures for determining the distribution of currents and voltage drops in combination circuits.

In most practical situations, you are given the total source voltage and the values of the individual resistances. The circuit in *Figure 6-6* illustrates the procedure for analyzing a combination circuit in terms of total resistance, total current, current through each resistor, and the voltage drop across each resistor.

Step 1 represents the only meaningful thing you can do first—combine parallel resistors R_2 and R_3 into an equivalent resistance, R_A. Step 2 continues the procedure by combining the values of R_1 and R_A to determine the total resistance of the circuit.

Once you have determined the total resistance of the circuit and replaced all resistors with a single equivalent value, you can use Ohm's law to calculate the total circuit current. This is shown as part of Step 3.

The equivalent circuit in Step 4 is identical to the one in Step 2. In this instance, however, you have more information to work with. The resistors are connected in series, so it follows that the same current—the total circuit current—flows through both of them. You can then apply Ohm's law to resistor R_1 to calculate the voltage drop across it. Likewise, Ohm's law provides the voltage drop across the equivalent resistance, R_A. Notice that the sum of those calculated values in Step 4 is very close to the total voltage.

The circuit in Step 5 is identical to the original circuit. It shows no equivalent resistances. At this point in the procedure, you know that the voltage across the parallel combination of R_2 and R_3 is 6.9 V. Now you can use Ohm's law to calculate the currents through each of them. The sum of those two currents is close to the value of total current.

The calculations in Step 5 complete the analysis of this circuit. Summarizing all of the information:

$E_T = 12$ V	(Given)	$R_1 = 100\ \Omega$	(Given)
$R_T = 233\ \Omega$	(Step 2)	$I_1 = 52$ mA	(Step 4)
$I_T = 52$ mA	(Step 3)	$E_1 = 5.2$ V	(Step 4)
$R_2 = 400\ \Omega$	(Given)	$R_3 = 200\ \Omega$	(Given)
$I_2 = 17$ mA	(Step 5)	$I_2 = 35$ mA	(Step 5)
$E_2 = 6.9$ V	(Step 4)	$E_2 = 6.9$ V	(Step 4)

COMBINED SERIES AND PARALLEL CIRCUITS

**Figure 6-6.
Complete Analysis,
Simple Combination
Circuit**

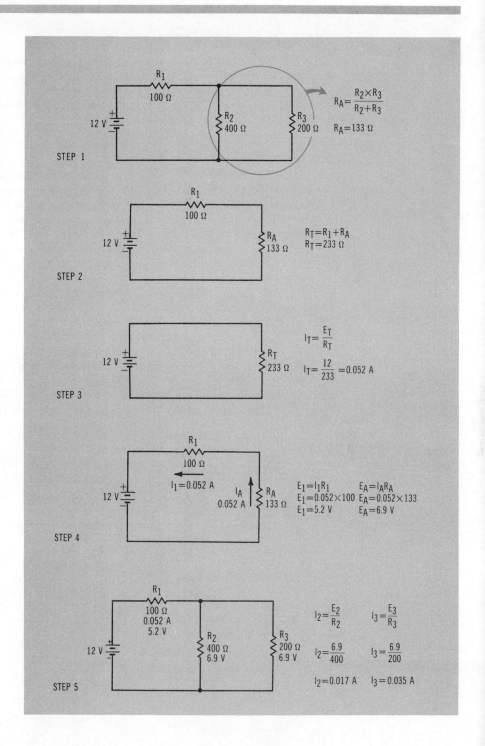

6 COMBINED SERIES AND PARALLEL CIRCUITS

KIRCHHOFF'S LAW

Kirchhoff's law defines the distribution of currents and voltages within an electrical circuit. It is used to check whether you have assigned the proper direction for current flow and to double-check your analysis of a combination circuit. Kirchhoff's law can also be helpful when you are given values other than total voltage and the resistance of the individual resistors.

Kirchhoff's Voltage Law

The point at which current enters a resistance is negative with respect to the point at which it leaves.

You have already seen how you can analyze a circuit in terms of the voltage drops across each resistor. And you have seen that you can determine the polarity of the voltage drop across a resistor: the point at which electron current enters a resistance is negative with respect to the point at which it leaves.

Upon determining the IR voltage drop and polarity across the resistors, you will find that the sum of the voltages around any loop in the circuit is equal to zero. This is an informal statement of Kirchhoff's voltage law. Before you are fully prepared to understand and use this law, you must understand exactly what summing the values in a loop means.

Figure 6-7 shows a simple series circuit composed of three resistors. The values of the resistors are not significant yet. More important are the IR voltage drops across the resistors, the source voltage, and the polarities of the voltages. The current flows in a counterclockwise direction. Electrons leave the negative terminal of the 8-V source, flow through the three resistors, and return to the positive terminal of the source. Notice that the IR drops are negative where current enters a resistor and positive where the current leaves the resistor.

**Figure 6-7.
Simple Kirchhoff Voltage Loop**

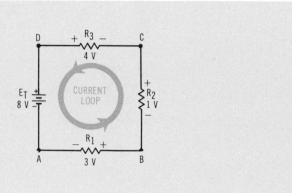

COMBINED SERIES AND PARALLEL CIRCUITS

According to Kirchhoff's voltage law, a loop is any complete path that begins at one point and returns to that point from the opposite direction. The loop, for example, can begin at Point A, run through R_3, R_2, and R_1, and then run through the voltage source (E_T) to return to Point A. You can actually begin the loop at any point you choose, just as long as you follow the same path and return to the starting point from the opposite direction.

If you sum all of the voltages along the path of a loop—paying attention to the polarity where you enter a component—the result is zero. Beginning from Point A in the figure, you first encounter a -3 V at R_3. Then the loop leads you through -1 V for R_2 and -4 V for R_1. But the loop is not complete at that point. It must also include the source, E_T. The first polarity you encounter at the battery is positive. So E_T adds a $+8$ V to the sequence of sums. The overall sequence, an example of Kirchhoff's voltage law, looks like this:

$$-3 -1 -4 +8 = 0$$

Kirchhoff's voltage law: The algebraic sum of voltages in a loop is equal to zero.

Summing minus and plus values in that fashion is called an "algebraic sum." Now we can formally state Kirchhoff's voltage law: the algebraic sum of voltages in a loop is zero. As long as the IR drops and polarities are correct, the algebraic sum of the voltages, including the source voltage, is zero. You can see that the law holds true, no matter where you begin the loop. And you can actually run the loop in the reverse direction and get the same result.

Figure 6-8 shows a combination circuit that has three different Kirchhoff loops. Loop 1 is a small loop that includes the IR drops across resistors R_1, R_2, and the source voltage. Loop 2 includes only IR drops—those across resistors R_2, R_3, and R_4. The third possible loop, Loop 3, includes the IR drops across resistors R_1, R_2, and R_3, as well as the source voltage. Notice that the algebraic sums of the voltages in all three loops add up to zero.

Kirchhoff's Current Law

You have just seen that Kirchhoff's voltage law makes a clear statement about the distribution of voltages in electrical circuits. There is also a current law—a form of Kirchhoff's Law that deals exclusively with the distribution of current in a circuit.

6
COMBINED SERIES AND PARALLEL CIRCUITS

**Figure 6-8.
Three Kirchhoff Voltage Loops**

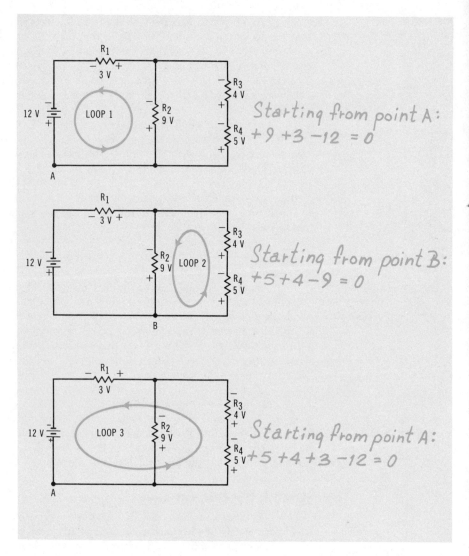

Kirchhoff's current law: the algebraic sum of currents entering and leaving a point in a circuit is equal to zero.

According to Kirchoff's current law, the algebraic sum of the currents entering and leaving any point in a circuit is equal to zero. *Figure 6-9* illustrates this basic idea. Current I_1 is entering the point, while currents I_2 and I_3 are leaving it. If you assign a plus value to currents entering a point and negative values to those leaving a point, you should see that their algebraic sum is zero.

COMBINED SERIES AND PARALLEL CIRCUITS

Figure 6-9.
Currents Entering and Leaving a Point

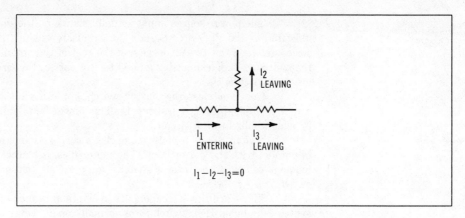

Suppose that I_1 is 5 A, and I_2 and I_3 are 2 A and 3 A, respectively. Because I_1 is entering the point, it has a value of +5; because I_2 and I_3 are leaving the point, they have values of −2 and −3. Doing an algebraic sum demonstrates Kirchhoff's current law:

$$+5 - 2 - 3 = 0$$

HOW SWITCHES ARE USED IN COMBINATION CIRCUITS

Switches are used in combination circuits to switch electrical power to various components. The circuit in *Figure 6-10* shows an arrangement of switches and load devices. It is typical of many kinds of machines used in modern industry and automated office equipment.

Figure 6-10.
Switch-Control Circuit

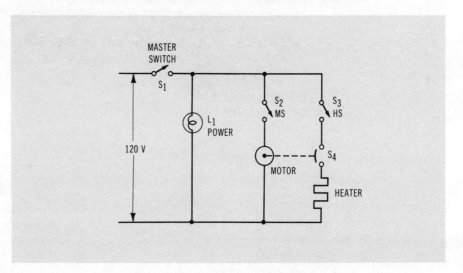

UNDERSTANDING ELECTRICITY AND ELECTRONICS CIRCUITS

6. COMBINED SERIES AND PARALLEL CIRCUITS

The power source is not actually shown on the diagram, but it is indicated as 120 V. Switch S_1, labeled the master switch, is connected in series between the 120-V source and the rest of the circuit. This means that the source voltage cannot be applied to any part of the circuit until the master switch is closed.

Once you close the master switch, you can see that the 120-V source is applied directly to lamp L_1. This power lamp indicates that power is being applied to the circuit.

Switch S_2 and a motor comprise a series circuit that is connected in parallel with the power lamp. The motor does not necessarily run the moment you close the master switch—you must also close S_2, the MS switch.

There are three components in the final branch of the circuit: switch S_3 (labeled HS), switch S_4 (a normally open "centrifugal switch" that closes only when the motor is running at full speed), and an electrical heating element. You can see that electrical power is applied to the heating element only when switches S_1, S_3 and S_4 are closed. And S_4 closes only when the motor is running at full speed.

WHAT HAVE WE LEARNED?

1. A combination circuit provides both series and parallel paths for current flow.
2. To find the total resistance of a combination circuit, begin combining resistances wherever you can, substitute equivalent resistance values, and continue in a systematic fashion until the circuit is reduced to a single equivalent resistance.
3. The point at which electron current enters a resistance is negative with respect to the point at which it leaves.
4. Kirchhoff's voltage law states that the algebraic sum of voltages in a complete loop is equal to zero.
5. Kirchhoff's current law states that the algebraic sum of currents entering and leaving a point in a circuit is equal to zero.

KEY WORDS

Centrifugal switch
Combination circuit

Quiz for Chapter 6

1. Which one of the following is a direct statement of Ohm's law?
   a. The total resistance of a series circuit is equal to the sum of the individual resistances.
   b. The total resistance of a parallel circuit is less than the value of the smallest resistance.
   c. The algebraic sum of voltages in a loop is equal to zero.
   d. The algebraic sum of currents entering and leaving a point is equal to zero.
   e. The voltage drop across a resistance is proportional to the value of resistance and the amount of current flowing through it.

2. Which one of the following is a statement of Kirchhoff's voltage law?
   a. The total voltage in a parallel circuit is less than the value of the smallest voltage.
   b. The algebraic sum of voltages in a loop is equal to zero.
   c. The algebraic sum of voltages entering and leaving a point is equal to zero.
   d. The voltage drop across a resistance is proportional to the value of resistance and the amount of current flowing through it.

3. Which one of the following is a statement of Kirchhoff's current law?
   a. The total current in a series circuit is equal to the sum of the individual currents.
   b. The total current in a parallel circuit is less than the value of the smallest current.
   c. The algebraic sum of currents in a loop is equal to zero.
   d. The algebraic sum of currents entering and leaving a point is equal to zero.
   e. The current through a resistance is proportional to the value of resistance and the IR drop across it.

4. What is the algebraic sum of this series of values: $+2$, -3, -6, $+4$?
   a. -15.
   b. -3.
   c. 0
   d. $+2$.
   e. $+3$.
   f. $+15$.

5. What is the total resistance of the circuit in *Figure 6-11*?
   a. 91 Ω.
   b. 340 Ω.
   c. 550 Ω.
   d. 600 Ω.
   e. 1100 Ω.

6. What is the total current for the circuit in *Figure 6-11*?
   a. 16 mA.
   b. 28 mA.
   c. 31 mA.
   d. 50 mA.
   e. 187 mA.

7. What is the IR drop and current for resistor R_1 in *Figure 6-11*?
   a. 1.7 V, 50 mA.
   b. 16 V, 16 mA.
   c. 5 V, 50 mA.
   d. 3 V, 30 mA.
   e. None of these.

6 COMBINED SERIES AND PARALLEL CIRCUITS

**Figure 6-11.
Circuit for Quiz
Questions 5–13**

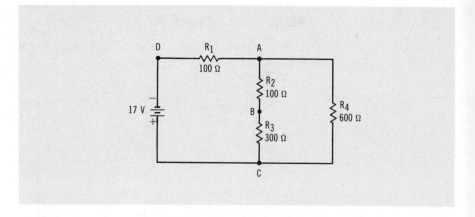

8. What is the IR drop and current for resistor R_2 in *Figure 6-11*?
   a. 1.7 V, 50 mA.
   b. 16 V, 16 mA.
   c. 5 V, 50 mA.
   d. 3 V, 30 mA.
   e. None of these.

9. What is the IR drop and current for resistor R_3 in *Figure 6-11*?
   a. 3 V, 10 mA.
   b. 3 V, 30 mA.
   c. 9 V, 30 mA.
   d. 12 V, 30 mA.
   e. 12 V, 40 mA.
   f. None of these.

10. What is the IR drop and current for resistor R_4 in *Figure 6-11*?
    a. 9 V, 15 mA.
    b. 12 V, 20 mA.
    c. 12 V, 30 mA.
    d. 14 V, 20 mA.
    e. 15 V, 25 mA.
    f. None of these.

11. Which one of the following statements is true for the circuit in *Figure 6-11*?
    a. Point A is positive with respect to Point B.
    b. Point A is negative with respect to Point B.
    c. Point A is negative with respect to Point D.
    d. Point A is positive with respect to Point C.

12. Which one of the following statements is true for the circuit in *Figure 6-11*?
    a. Point B is positive with respect to Point A.
    b. Point B is negative with respect to Point A.
    c. Point B is negative with respect to Point D.
    d. Point B is positive with respect to Point C.

13. Which one of the following statements is true for the circuit in *Figure 6-11*?
    a. Two electron currents are entering Point C and one is leaving.
    b. Two electron currents are leaving Point C and one is entering.
    c. Two electron currents are entering Point A and one is leaving.
    d. Two electron currents are entering Point B and none is leaving.

Understanding Magnetism

ABOUT THIS CHAPTER

This chapter explains the principles of magnetism. When you complete the chapter, you should be able to visualize and describe the basic principles and terminology of magnetism.

FROM THE BEGINNING

In a district in Asia Minor known as Magnesia, the ancient Greeks noticed that a lead-colored stone attracted small particles of iron ore. The stone, shown in *Figure 7-1a*, was called loadstone, meaning "leading stone."

Figure 7-1. Natural Magnet

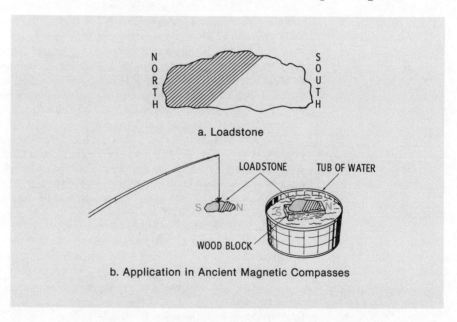

a. Loadstone

b. Application in Ancient Magnetic Compasses

The Chinese learned that when the loadstone is suspended or floated on a liquid, one end of the stone always pointed in a given direction. They used loadstone as the main working element in their navigational compasses, as shown in *Figure 7-1b*.

7 UNDERSTANDING MAGNETISM

Loadstone (also spelled lodestone) is a natural magnet because it possesses magnetic properties in its natural state. Although loadstone still exists in nature, today's magnets are artificial. Magnets have a variety of applications in modern technology, including producing most electrical energy.

Dictionaries define magnetism as "a peculiar property possessed by certain materials by which they can naturally repel or attract one another according to determined laws." In order to understand magnetism, it is necessary to study the properties and laws because magnetism, like electricity, is actually an invisible force, although you can observe its effects on other materials.

Artificial magnets are much stronger than natural loadstone. Iron, cobalt, and nickel are used in the manufacture of these magnets. Soft iron is easy to magnetize, but it loses its magnetic properties almost immediately after the magnetizing force is removed. Steel is harder to magnetize, but it holds its magnetism over a greater period of time after the magnetizing force is removed.

MAGNETIC CHARACTERISTICS

Modern versions of the Chinese loadstone compass use artificial magnets. The principle of operation is the same in both instances, however. Today we know that any magnetic or magnetized material, when suspended in or floated on a liquid so that it can rotate freely, aligns itself with the earth's own natural magnetic field. The end of the compass magnet that points toward the earth's magnetic north pole is called the "north-seeking" pole. The opposite end of the compass magnet is called the "south-seeking" pole. This early tradition gives us today's names for the two poles of a magnet: the north (N) and south (S) poles.

Iron and certain iron alloys are useful for explaining magnetic properties and effects because magnetism is more pronounced in them. Notice in *Figure 7-2* that the majority of electrons in orbit about the nucleus appear to be traveling in the same direction. This is a clue as to why certain materials are easy to magnetize while others are almost impossible to magnetize. If you could take a close look at the atoms in materials that cannot be magnetized, you would see that the electrons appear to be spinning in different directions; thus, they cancel each other's magnetic effects and prevent the development of any significant external magnetic field.

Unlike magnetic poles attract one another, whereas like poles repel.

You have learned that any magnet or magnetized material has north and south poles. One of the most important characteristics possessed by magnets is that like poles (N and N, or S and S) repel one another, while unlike poles (N and S) attract each other.

UNDERSTANDING MAGNETISM

**Figure 7-2.
Atom of Iron**

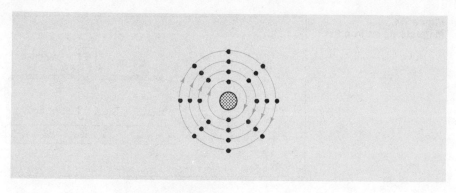

Molecular Theory of Magnetism

If you were able to view the molecules inside a block of unmagnetized iron, you would see a total disarrangement of the molecules, as shown in *Figure 7-3*. Each molecule within a bar of iron has its own north and south poles. Although the magnetic strength of a single molecule is very weak, there are many millions of molecules in a very small piece of metal. When magnetically aligned in the same direction, they can develop strong magnetic effects. This is known as the "molecular theory of magnetism."

Most of the magnetic poles of molecules in a magnet are aligned in the same direction. The magnetic poles of the molecules of a non-magnetized material are aligned at random.

**Figure 7-3.
Molecules Inside a Material**

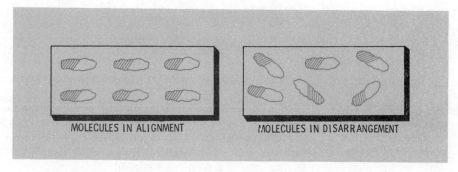

MOLECULES IN ALIGNMENT   MOLECULES IN DISARRANGEMENT

To magnetize a bar of iron, stroke the bar with a material that is known to be a magnet. Suppose that you choose to apply the north pole of the magnet to the bar of iron and strike the iron bar from left to right as shown in *Figure 7-4*. Note that the same pole of the magnet is always applied to the iron bar and that the stroking action is always in the same direction. Make sure the magnet is lifted away from the iron bar at the end of each left-to-right stroke.

UNDERSTANDING ELECTRICITY AND ELECTRONICS CIRCUITS

7 UNDERSTANDING MAGNETISM

Figure 7-4.
Magnetizing an Iron Bar

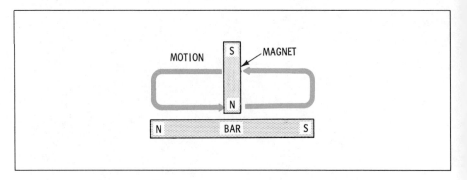

Another way to magnetize an iron bar is to apply a magnet at the center of the bar and stroke in one direction. After half the bar is magnetized, reverse the magnet and, again starting at the center, stroke the iron bar in the opposite direction. See the procedure as illustrated in *Figure 7-5.*

Figure 7-5.
Alternate Method for Magnetizing an Iron Bar

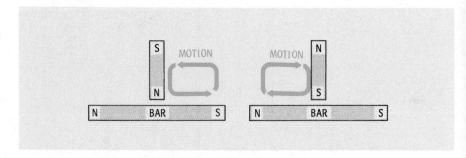

Retentivity is the property of a material to remain magnetized when a magnetizing force is removed.

A steel bar can be magnetized in exactly the same way. Steel requires a greater force to align its molecules and, therefore, takes longer to magnetize. However, steel retains its magnetic properties for a longer time than iron and it is said to have a high "retentivity" (the property of any material to remain magnetized).

Magnetic Field

Magnetic lines of force leave the north pole, pass through a magnetic circuit, and return to the south pole.

All magnets have invisible lines of force surrounding them. These lines leave the north pole of a magnet, form a loop, and return to the south pole of the magnet. The loops run parallel to each other; they never cross nor link up with one another. See the examples in *Figure 7-6.*

The lines formed by magnetic loops around a magnet are called "magnetic lines of force." The area occupied by these lines of force is called the "magnetic field."

104 UNDERSTANDING ELECTRICITY AND ELECTRONICS CIRCUITS

UNDERSTANDING MAGNETISM

Figure 7-6.
Paths of Magnetic Lines of Force

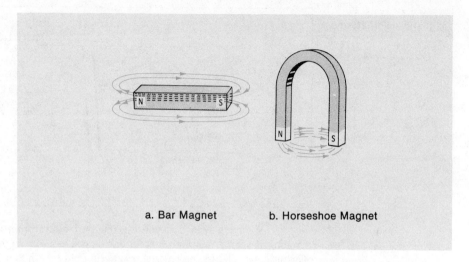

a. Bar Magnet   b. Horseshoe Magnet

The magnetic field is the induced energy surrounding the magnet or the space through which the influence of these magnetic lines of force can be measured. The strength of the magnetic field is measured by determining the number of magnetic lines of force per unit area surrounding the magnet. Since these lines are invisible, how can their total number be counted or pattern they form be determined? A simple experiment that you may wish to perform answers these questions. It allows you to see the lines of force for yourself.

In order to perform the experiment you need:

1. A bar or horseshoe magnet
2. A piece of glass or clear plastic about 12 inches square
3. A small container of iron filings.

Place the glass or plastic sheet over the magnet and sprinkle a small amount of iron filings (about a thimble full) over the magnet area on the surface of the sheet. Tap the sheet and notice how the iron filings form a definite pattern similar to that shown in *Figure 7-7*. Recall that the lines of force are invisible but that you can observe their effects. Notice the heavy concentration of iron filings near the poles of the magnet.

Whenever a piece of iron or steel is placed into a magnetic field, it assumes the properties of that magnetic field. You probably know the danger to your mechanical wristwatch if you wear it when working near strong magnets. The gears and other small metallic parts can become magnetized, thus binding up the works so that the watch no longer can run. Many mechanical watches made today are said to be non-magnetic. This does not mean you can lay your watch on a strong magnet without ruining it. It does mean, however, that the watch is shielded from magnetic fields encountered in everyday situations. (Digital watches are not affected by magnetic fields because they have no metallic moving parts.)

**Figure 7-7.
Magnetic-Field Patterns**

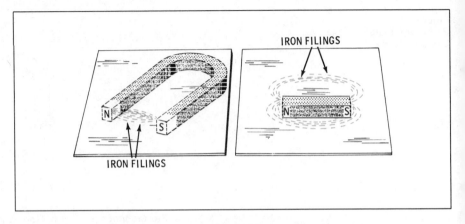

We have seen that you can magnetize an iron or steel bar by stroking it with a magnet. A steel bar or rod can also be magnetized by placing it parallel to the earth's magnetic field (north and south) and striking it several sharp blows with a hammer. The force from these blows causes the molecules in the bar or rod to change positions and to align themselves with the earth's magnetic field.

Commercially available magnets are manufactured by placing the magnetic material into a strong magnetic field created by an electromagnet. Materials can also be demagnetized by placing them into a magnetic field that changes polarity very rapidly. Rapidly changing magnetic fields are easily obtained from electromagnets. We will examine electromagnets in detail in Chapter 8.

The path of magnetic lines of force can be controlled. The lines of force concentrated at the poles of a magnet are much closer together than those surrounding the middle of the magnet. This is true because magnetic lines of force always take the path of least opposition. Iron or steel offers less opposition to these lines of force than air or other non-magnetic materials. This principle can be used to advantage. As shown in *Figure 7-8*, for example, if an iron or steel ring is placed around a mechanical wristwatch, magnetic lines of force follow the path through the ring and do not pass through the watch. This method of diverting magnetic lines of force is called "magnetic shielding."

Demagnetizing Metals

You can also demagnetize metals. When a steel screwdriver becomes magnetized, it tends to hold steel screws and bolts at odd angles that make it difficult to use the tool. Demagnetizing a screwdriver is a matter of striking it on a hard surface a few times. Providing its magnetic field is rather weak, this striking action will rearrange the molecules in a more random pattern, reducing the magnetic effect. Be sure the screwdriver is not aligned with the earth's magnetic field as you strike it on the hard surface, however.

**Figure 7-8.
Magnetic Shielding**

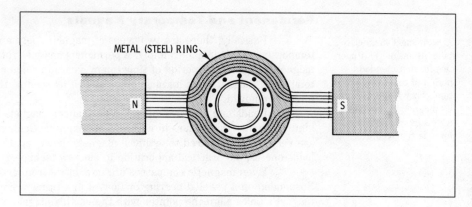

Heating, as well as jarring, reduces the magnetism of any material. When iron is heated to about 770°C, it can no longer be magnetized or hold a magnetic field. Heating a material accelerates the movement of its molecules, causing the molecules to assume a random alignment.

Magnetic Poles

The minimum number of poles a magnet can have is two—a north and south pole. But it is possible for a magnet to have more than two poles. The poles between the ends of a magnet are called "consequent poles." Notice in *Figure 7-9* that there are magnetic fields existing between the end and consequent poles. The magnetic lines of force for all poles of a magnet leave a north pole and enter a south pole.

**Figure 7-9.
Consequent and End Poles**

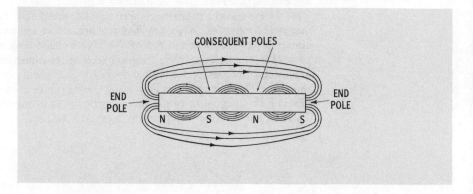

7 UNDERSTANDING MAGNETISM

Permanent and Temporary Magnets

A permanent magnet retains its magnetic properties for an extended period of time, whereas a temporary magnet loses its magnetic properties when the magnetizing force is removed.

Basically, there are two types of magnets: permanent and temporary. As their names imply, a permanent magnet retains its magnetism for a long period of time (many years in some cases), while a temporary magnet loses its magnetism almost as soon as the magnetizing force is removed.

Loadstone is an example of a permanent magnet. It is the only kind of magnet that occurs in nature. The magnetic fields around loadstones are very weak compared to artificial magnets, however. For that reason, loadstone has no practical application in modern technology.

Even magnetic compasses are now made from artificial magnets. Suspended on a jeweled bearing or floated in a transparent fluid, the magnet's poles align the pointer with the earth's magnetic poles. But you have just learned that iron and steel materials can distort magnetic fields. That creates something of a problem when attempting to operate an ordinary compass inside an automobile, for instance. The car's steel structure distorts the earth's magnetic fields and causes the compass to give unreliable readings.

Compasses that are intended to be used in automobiles or other steel enclosures use small bar magnets—compensating magnets—to counteract the effect of the steel on the earth's magnetic field. Once this kind of compass is fixed in place, you can adjust the compensating magnets to get reliable compass readings.

Permanent horseshoe magnets are used in most electrical meters, including voltmeters, ammeters, ohmmeters, and wattmeters. By bringing the two end poles close together in the horseshoe shape, the lines of force are concentrated and provide a strong magnetic field.

Permanent magnets require proper care if they are to retain their magnetic properties. When bar magnets are not in use, they should be stored in pairs as shown in *Figure 7-10*. Notice that they are aligned parallel to one another, with their unlike poles touching one another. When a horseshoe magnet is not in use, a soft iron bar should be placed across the poles. This bar provides a path for the magnetic lines of force, thereby enabling the magnet to retain its properties over a longer period of time. The iron bar used for this purpose is called a keeper.

**Figure 7-10.
Storing Magnets**

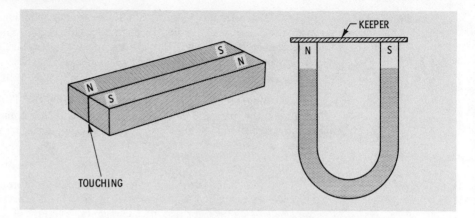

In special cases, it is important for the magnet to maintain a specific magnetic force over a long period of time. When necessary, the magnets are placed in an oven and subjected to controlled temperature changes and vibrations. This process, called "aging," causes the strength of the magnet to remain nearly constant for a long period of time.

BASIC UNITS OF MAGNETISM

In our study of electrical circuits, we saw that electromotive force (voltage) causes current to flow and that the flow of that current is limited by resistance. There is also a magnetic circuit in which the magnetic lines of force form closed, or "flux," loops. The force that produces flux loops is called "magnetomotive force (mmf)." The opposition to the flux loops is the property of "reluctance."

Notice the similarity between electric and magnetic circuits. In a magnetic circuit, the magnetic lines of force always take the path of least reluctance. This is why the magnetic lines follow a steel ring around a watch as described earlier: the steel ring offers less reluctance to the lines of force than the air and the non-magnetic material of the watch.

Magnetic Flux

An expression for determining the amount of flux present in a magnetic circuit is:

$$\text{flux} = \frac{\text{magnetomotive force}}{\text{reluctance}}$$

Flux varies directly with the magnetomotive force and inversely with the reluctance. This is the Ohm's law expression for magnetic circuits. Compare the equation with its electric counterpart.

$$\text{current} = \frac{\text{electromotive force}}{\text{resistance}}$$

7 UNDERSTANDING MAGNETISM

Magnetic flux is a measurement of the number of lines of force in a magnetic field.

Magnetic flux is the total number of magnetic lines existing in a magnetic circuit or extending through a specific region. The units of measurement for magnetic flux are the maxwell or weber, depending on the system of units you are using.

Magnetomotive Force

Magnetomotive force is the physical force that creates and maintains the magnetic field around a magnet. It is the magnetic counterpart of voltage in electrical circuits. The unit of measurement varies, depending on the units used for other terms.

Reluctance

Reluctance is the resistance to the build-up of magnetic fields in response to an applied magnetomotive force. This is a property of the materials being used for making the magnet. The unit of measurement varies with the units used for other magnetic terms.

Flux Density

Flux density is a measure of the number of lines of magnetic force per unit area.

The concentration of magnetic lines determines the "flux density." Generally speaking, flux density is the number of magnetic lines per unit of area. The unit of measurement is either the gauss or tesla. One gauss is defined as one line of force, or maxwell, per square centimeter. One tesla equals one weber per square meter. Flux density can be expressed as follows:

$$\text{flux density} = \frac{\text{number of lines of force}}{\text{area}}$$

For a given area, increasing the number of lines of force increases the flux density of the magnetic field. Or, if you maintain the same number of lines of force but increase the area, the flux density decreases.

The flux density between the poles of a horseshoe magnet is directly proportional to the area of the air gap between the poles. The force of attraction or repulsion between the poles varies directly with the strength of the poles and inversely with the square of the distance separating them.

Permeance

Permeance is a measure of how easily magnetic fields can be concentrated within a material.

Certain materials have more opposition, or reluctance, to magnetic lines of force than others. It is therefore true that some materials allow magnetic lines of force to pass more easily than others. The ease with which magnetic lines of force pass through a material is known as "permeance," the reciprocal of reluctance.

A substance that allows the magnetic flux to pass with little or no opposition is said to have a high permeability. Iron, for example, has a high permeability. High-permeability materials can be easily magnetized, but they do not retain their magnetism. Permeability varies with the intensity

UNDERSTANDING MAGNETISM

of the magnetic field in which the material is located. Another way to determine the flux in a material is to multiply the magnetomotive force by the permeance.

$$\text{flux} = \text{mmf} \times \text{permeance}$$

Levels of Permeability

Diamagnetic materials cannot be magnetized under normal conditions. Paramagnetic materials can be magnetized to some extent. Ferromagnetic materials can be easily magnetized.

Not all materials can be magnetized. Three classifications according to permeability describe magnetic capability: diamagnetic, paramagnetic, and ferromagnetic. Diamagnetic materials are those that normally cannot be magnetized; they have a permeability of less than 1. Paramagnetic materials, such as aluminum and platinum, can be magnetized, but only with great difficulty; they have a permeability that is slightly greater than 1. Ferromagnetic materials are those that are relatively easy to magnetize; their permeability is quite high. Ferromagnetic materials include iron, cobalt, nickel, silicon steel, and cast steel.

WHAT HAVE WE LEARNED?

1. Magnets have two poles that are designated north and south.
2. Unlike magnetic poles attract one another; like poles repel.
3. The magnetic poles of molecules in a non-magnetized material are arranged at random. The poles in a magnetized material are aligned in the same north-south direction.
4. You can magnetize an iron bar by stroking it with a magnet.
5. Commercial magnets are made by placing a soft iron bar within the intense magnetic field of an electromagnet.
6. Materials can be demagnetized by striking them sharply on a hard surface or by heating them.
7. Retentivity is the property of a material to remain magnetized when a magnetizing force is removed.
8. Magnetic lines of force leave the north pole, pass through a magnetic circuit, and return to the south pole.
9. A permanent magnet retains its magnetic properties for an extended period of time, whereas a temporary magnet loses its magnetic properties as soon as the magnetizing force is removed.
10. Reluctance is the property of a material that opposes the concentration of magnetic lines of force. Reluctance is to magnetic circuits as resistance is to electrical circuits.
11. Magnetomotive force (mmf) is the force that produces the flux lines around a magnet. Magnetomotive force is to magnetic circuits as voltage is to electrical circuits.
12. Magnetic flux is the total number of magnetic lines in a magnetic circuit. The unit of measurement is the maxwell or weber.
13. Flux density is a measure of the number of lines of magnetic force per unit area. The unit of measurement is the tesla or gauss.

14. Permeance is a measure of how easily magnetic fields can be concentrated within a material.
15. Diamagnetic materials normally cannot be magnetized, paramagnetic materials can be magnetized with difficulty, and ferromagnetic materials can be easily magnetized.

KEY WORDS

Bar magnet
Consequent magnetic poles
Diamagnetic material
Ferromagnetic material
Flux density
Flux lines
Horseshoe magnet
Keeper
Loadstone
Magnet
Magnetic field
Magnetic lines of force
Magnetic poles
Magnetic shielding
Magnetism
Magnetomotive force (mmf)
Maxwell
Paramagnetic material
Permanent magnet
Permeability
Permeance
Reluctance
Retentivity
Temporary magnet
Weber

Quiz for Chapter 7

1. A loadstone is:
   a. a natural magnet.
   b. an artificial magnet.
   c. a temporary magnet.
   d. a paramagnetic alloy.

2. The north pole of a magnet is the one:
   a. that has its molecules arranged in a random pattern.
   b. that would be attracted to the earth's north magnetic pole.
   c. that would be attracted to the earth's south magnetic pole.
   d. that would be repelled by the south pole of another magnet.

3. The south pole of a magnet is the one:
   a. that has its molecules arranged in a random pattern.
   b. that would be attracted to the earth's north magnetic pole.
   c. that would be attracted to the earth's south magnetic pole.
   d. that would be repelled by the north pole of another magnet.

4. Which one of the following statements is true?
   a. Like magnetic poles attract; unlike poles repel.
   b. Like magnetic poles repel; unlike poles attract.
   c. Whether magnetic poles repel or attract depends on their relative strength rather than their polarities.

5. Which one of the following properties of magnetic materials determines how well a material can hold a magnetic field after the magnetizing force is removed?
   a. Reluctance.
   b. Permeability.
   c. Retentivity.
   d. Permeance.
   e. None of these.

6. Which one of the following properties of a magnetic circuit can be directly compared to resistance in an electrical circuit?
   a. Reluctance.
   b. Permeability.
   c. Retentivity.
   d. Permeance.
   e. There is no valid comparison.

7. Which one of the following kinds of materials has the highest permeability?
   a. A diamagnetic material.
   b. A ferromagnetic material.
   c. A paramagnetic material.

8. Which one of the following kinds of materials has the lowest permeability?
   a. A diamagnetic material.
   b. A ferromagnetic material.
   c. A paramagnetic material.

9. Which one of the following properties of a magnetic circuit can be directly compared to voltage in an electrical circuit?
   a. Number of lines of force.
   b. Flux density.
   c. Magnetomotive force.
   d. Reluctance.

10. Which one of the following properties of a magnetic circuit can be directly compared to current in an electrical circuit?
    a. Number of lines of force.
    b. Flux density.
    c. Magnetomotive force.
    d. Reluctance.

11. Which one of the following phrases best describes flux density?
    a. Total number of lines of magnetic force.
    b. The rate of change of the number of lines of magnetic force.
    c. The opposition to a build-up of magnetic fields.
    d. The force that maintains lines of force.
    e. The concentration of lines of magnetic force.

8

Understanding Electromagnetism

ABOUT THIS CHAPTER

This chapter extends our examination of magnetism to include the important effects of electromagnetism. You will learn about the basic principles and terminology of electromagnets and discover how the same principles apply to the operation of electric motors and generators.

ELECTROMAGNETS

The bar and horseshoe magnets described in the preceding chapter are regarded as permanent magnets. A second classification of magnets could be called temporary magnets. One truly useful kind of temporary magnet is an electromagnet.

The magnetic field around an electromagnet is created by current flowing through a coil of wire. An electromagnet retains its magnetic field only as long as current is flowing through the coil.

An electromagnet is "a bar of soft iron that will become a temporary magnet if an electrical current is caused to pass through a wire that is coiled around it." This effect was described in 1828 by the Danish scientist, Hans C. Oersted. Oersted found that a compass needle placed near a wire would deflect when current passed through the wire. Further experimentation showed that current flowing through a conductor creates a magnetic field around the conductor. As shown in *Figure 8-1*, this magnetic field is composed of lines of flux that encircle the conductor at right angles to the flow of current. The flux lines are uniformly spaced along the length of the conductor.

**Figure 8-1.
Flux Lines Around Current-Carrying Conductor**

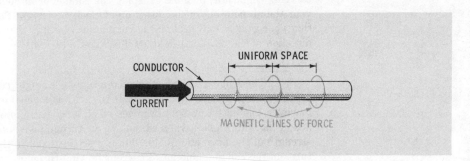

8 UNDERSTANDING ELECTROMAGNETISM

ELECTROMAGNETIC STRENGTH AND POLARITY
Strength

Reversing the direction of current through an electromagnet reverses its magnetic polarity.

Current flowing through a conductor creates a magnetic field. The magnetic field around a straight conductor is not very strong, but it exhibits the same properties as the bar or horseshoe magnet discussed previously. If a compass is placed near the conductor, the compass needle is deflected at right angles to the current-carrying conductor. The compass also indicates that the magnetic field around the conductor is polarized—if the current is reversed through the conductor, the position of the compass needle also reverses. As shown in *Figure 8-2*, the strength of the magnetic field around a current-carrying conductor decreases with distance from the conductor.

**Figure 8-2.
Magnetic Field Strength**

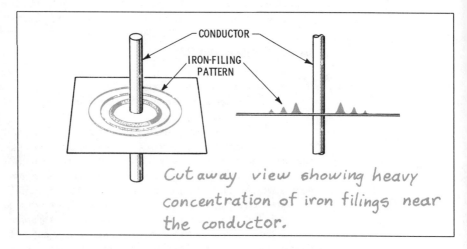

To determine the direction of the magnetic field around a current-carrying conductor, imagine grasping the conductor in your left hand, with your thumb pointing in the direction of electron current flow. The direction that your fingers curl around the conductor indicates the north-to-south direction of the magnetic field, as illustrated in *Figure 8-3*.

Increasing Strength

The magnetic field developed around a straight conductor is seldom strong enough to be useful. If the wire is formed into a coil, however, the magnetic field becomes much stronger. *Figure 8-4* shows the action that takes place when current flows through a coil of wire. The strength of the magnetic field is directly proportional to the number of turns in the coil and amount of current flowing through it. Increasing the number of turns increases the strength of the electromagnetic effect. Likewise, increasing the current flowing through the conductor increases the strength of the magnetic field.

UNDERSTANDING ELECTROMAGNETISM

Figure 8-3.
Left-Hand Rule

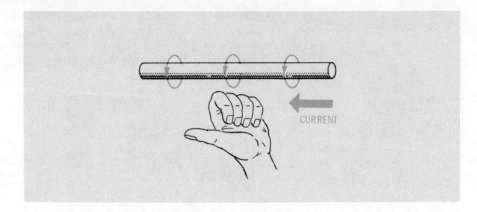

Figure 8-4.
Current Flow Through Coil

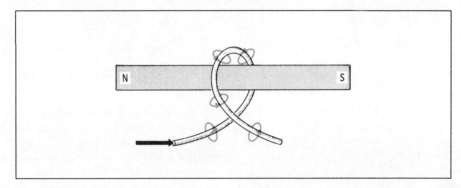

The coil of wire for an electromagnet is often wound around an iron core in order to concentrate the lines of force in a smaller area, thereby creating a stronger magnetic field.

The strength of an electromagnet is specified in terms of ampere-turns.

We saw in Chapter 7 that flux density is much greater in a block of iron than in the air. Therefore, if an iron core is added to the current-carrying coil of an electromagnet, the magnetic loops will concentrate through the core, increasing the flux density and, therefore, the strength of the electromagnet.

Measuring Strength

As we have seen, the two main factors that determine the magnetomotive force of an electromagnet are the current and the number of turns in the coil. The magnetic field can be varied by altering either of these factors. The product of current times number of turns for an electromagnet is called "ampere-turns." An electromagnet having 200 turns of wire through which 1 A of current is flowing will have a magnetomotive force that is the same as an electromagnet having a 10-turn coil with 20 A flowing through it.

Polarity

The magnetic lines of force generated by a coil of wire create a set of end poles similar to those of a permanent bar magnet. The polarity of the end poles depends on the direction the wire is coiled (clockwise or counterclockwise) and the direction of the current flowing through the wire.

UNDERSTANDING ELECTRICITY AND ELECTRONICS CIRCUITS

8
UNDERSTANDING ELECTROMAGNETISM

As indicated in *Figure 8-5*, you can also use the left-hand rule to determine the overall polarity of an electromagnet. Imagine holding the coil in your left hand, with your fingers curved in the direction of current flow through the coil of wire. Your thumb then points in the direction of the north end pole of the electromagnet.

**Figure 8-5.
Left-Hand Rule for Polarity**

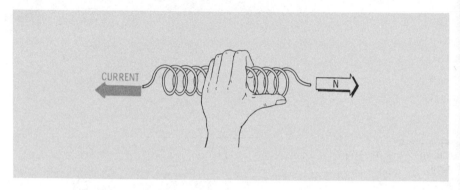

RESIDUAL MAGNETISM AND HYSTERESIS
Residual Magnetism

Residual magnetism is what remains in the core material of an electromagnet after current is removed from the coil.

When an electromagnet is de-energized, the magnetic field around the coil collapses, but a slight amount of magnetism remains in the core material. This is called "residual magnetism." Turning off the current does not eliminate the residual magnetism. You might think that turning the current on in the opposite direction would eliminate the residual magnetism. That's a good idea, because the opposing magnetic field from the coil could wipe out that remaining in the core. However, there is always the chance that you would overdo the procedure, finishing with some residual magnetism of the opposite polarity. This is always the case where you are using an electromagnet in a circuit that changes polarity frequently, 120 times a second for example.

If the current is reversed in an electromagnet (perhaps many times per second), the magnetic field and direction of polarization will also reverse. If the core material possesses any residual magnetism, the residual magnetism must be overcome before the core can be magnetized in the reverse direction.

Hysteresis

Hysteresis losses, in the form of heat, are especially apparent when the current changes polarity many times per second.

The discrepancy between the amount and direction of current flowing in the coil of an electromagnet and the amount and direction of the magnetic field in the core material is called "hysteresis." If the current through the coil is reversed frequently, a considerable amount of energy is lost in the form of heat. This is called "hysteresis loss."

UNDERSTANDING ELECTROMAGNETISM

When a further increase in current through an electromagnet causes no further increase in the magnetic field, the core is saturated.

The magnetic field around an electromagnet builds up as the current through the coil increases. There is a point, though, where the core material cannot accommodate additional lines of flux. When this point is reached, the core is said to be saturated. It is difficult to increase the flux density beyond the point of saturation, even by increasing the amount of current.

APPLICATIONS OF ELECTROMAGNETS

Solenoids

A coil wound in the shape of a cylinder is called a solenoid. A solenoid is often provided with a movable iron core, or plunger. In this arrangement, illustrated in *Figure 8-6*, the iron core is pulled into the coil when current flows through the turns of wire. Thus, the core can be used to move a device mechanically.

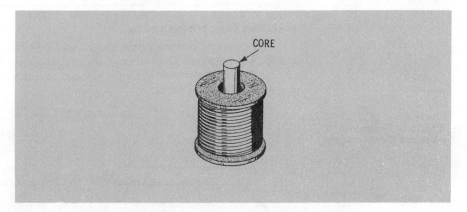

Figure 8-6. Solenoid

Solenoids are commonly used in relays, circuit breakers, and electrically operated hydraulic valves.

Toroids

Another type of coil that is used in a number of applications is called a toroid. As shown in *Figure 8-7*, a toroid has a ring-shaped core material on which the turns of wire are wound to form a complete circuit. This design concentrates all the lines of force inside the ring. With all lines of force thus confined to the ring, a toroid shows no external magnetic field.

**Figure 8-7.
Toroid**

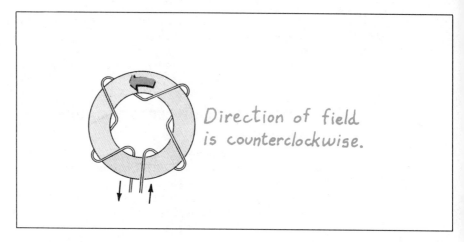

Polarized Electromagnets

A polarized electromagnet has a permanent magnet as its core material. When current flows through the coil, as shown in *Figure 8-8*, the magnetic field from the electromagnet can add to, cancel, or subtract from the permanent-magnet field.

**Figure 8-8.
Effects of Current in
Polarized Electromagnet**

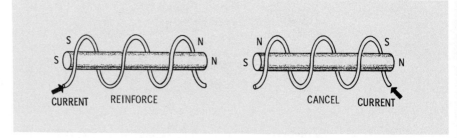

Figure 8-9 shows polarized electromagnets in a common doorbell appliance. The clapper arm, which extends down between the electromagnets, is attached to a permanent magnet. The permanent magnet holds the clapper arm in a neutral position between the bells and provides the clapper with a specific magnetic polarity. When current flows through the electromagnets, they generate magnetic fields of opposite polarity. The permanent magnet attached to the clapper arm is thus attracted to one electromagnet and repelled by the other and the clapper strikes the bell attached to the attracting electromagnet. When the current is reversed, the clapper moves in the direction that causes it to strike the other bell. Alternating the direction of current causes the clapper to move back and forth between the electromagnets, creating the familiar bell sound.

**Figure 8-9.
Polarized Electromagnet
in Doorbell**

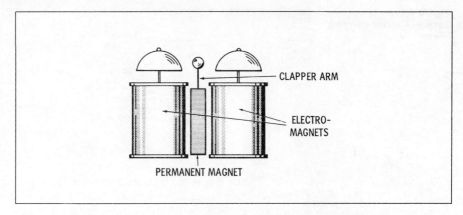

Inexpensive air pumps for fish aquariums work according to the same general principle, but the permanent magnet is attached to a flexible diaphragm. As the current switches polarity (120 times per second), the diaphragm creates enough pulsating air pressure to force air bubbles through the water in the aquarium.

You might ask why the permanent magnet in a polarized electromagnet assembly does not become demagnetized when the electromagnetic field opposes it. Once a permanent magnet becomes magnetized, a strong opposing magnetic force is required to disarrange its molecules. An electromagnetic field might be strong enough to do this if it remained for a very long period of time. However, the current through the electromagnets in such devices only lasts for brief periods of time.

ELECTROMAGNETISM IN MOTORS, GENERATORS, AND METERS

When current passes through a pair of parallel wires, the magnetic fields around those wires interact. If the currents flow in the same direction, the magnetic fields between them are in opposite directions, creating a force that tends to pull the wires together. If the currents flow in the opposite direction, the magnetic fields between the wires are in the same direction and tend to move the wires apart.

Figure 8-10 illustrates this situation. Notice that a dot indicates current flowing "out of the page," while a cross indicates current flowing "into the page."

**Figure 8-10.
Parallel Current-carrying Wires**

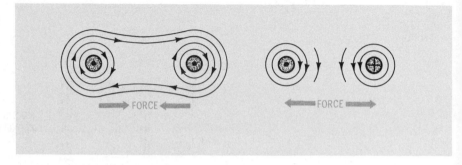

The principle of magnetic attraction and repulsion between current-carrying wires is the principle behind the operation of electric motors and generators. Electric motors are used to provide a mechanical power output from an electrical input. Generators provide an electrical power output from a mechanical input.

Figure 8-11 shows a current-carrying conductor that is located within the magnetic field of a permanent magnet. The current is shown flowing out of the page, generating a clockwise magnetic field. At the top of the conductor, its magnetic field and the field from the permanent magnet run in the same direction. As a result of that interaction, the conductor will be repelled by the upper portion of the permanent magnetic field. The situation is just the reverse at the lower end of the conductor. The two magnetic fields run in opposite directions, tending to attract the conductor downward. The two interactions complement one another—the upper portion of the permanent magnetic field pushes the conductor downward, and the lower portion of the permanent magnetic field pulls at the conductor. The combined force moves the conductor downward through the permanent magnetic field.

**Figure 8-11.
Current-carrying Conductor in Permanent Magnetic Field**

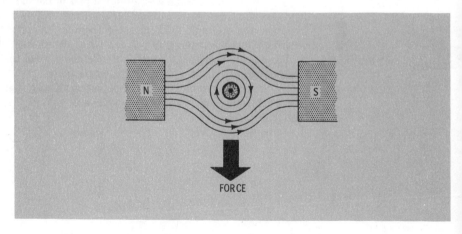

UNDERSTANDING ELECTROMAGNETISM

Normally the polarity of a permanent magnet remains unchanged. However, you can easily change the magnetic polarity of a conductor by reversing the direction of current flowing through it. What would happen if you changed the direction of current through the conductor in *Figure 8-11*? That action would create a circular counterclockwise magnetic field. The lines above the conductor would run in opposite directions, and those below the conductor would run in the same direction. The overall effect would be a force that moves the conductor upward.

There is another method for determining the direction of force generated by the interaction of the magnetic fields generated by a permanent magnet and a current-carrying conductor. This is the right-hand rule illustrated in *Figure 8-12*.

**Figure 8-12.
Right-Hand Rule for Motors**

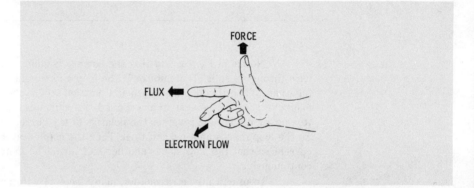

Arrange the thumb, index finger, and middle finger of your right hand as shown in the figure. Point the index finger in the direction of magnetic flux (the field from the permanent magnet) and the middle finger in the direction of current flow through the conductor. The thumb then indicates the direction of force exerted on the conductor.

Motors

A simple motor is shown in *Figure 8-13*. When current from the battery flows through the wire loop, it generates a magnetic field that interacts with the lines of flux from the permanent magnet. Using the right-hand rule, you can see that the interaction of fields will force the left side of the loop downward—electron flow in that part of the wire is into the loop, the flux lines run right to left, and the resulting force is downward. Current is flowing out of the loop at its right side. Applying the right-hand rule to that situation shows an upward force. Putting the information together, the loop makes at least a half turn in the counterclockwise direction. Reverse the polarity of the battery and you can bet the loop will turn clockwise.

8 UNDERSTANDING ELECTROMAGNETISM

**Figure 8-13
Simple Electric Motor**

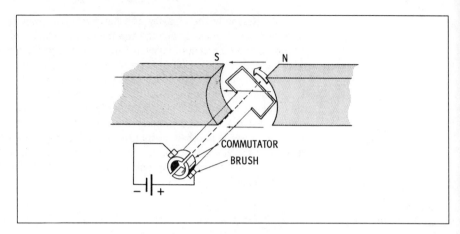

Commutator, Armature, and Brushes

The commutator forms the electrical connection between electrical source and movable conductor.

Notice in *Figure 8-13* that the battery is connected to the loop of wire through a split "commutator." The loop and commutator rotate together. When the loop has reached a position where the opposing magnetic poles of the system are aligned, the commutator will have rotated to a position where it reverses the polarity of the current through the loop. So the loop makes another half turn. Then the commutator reverses the current so that the loop makes another half turn. The motor runs continuously.

The movable coil of wire is the armature.

In practical electric motors, many loops of wire are wound around a core. This assembly is called an "armature." Each loop is connected to a commutator segment that makes contact with a set of brushes as the armature rotates. The use of many loops provides smoother operation and far more mechanical force than a single loop. An end view of an armature is shown in *Figure 8-14*.

**Figure 8-14.
Armature**

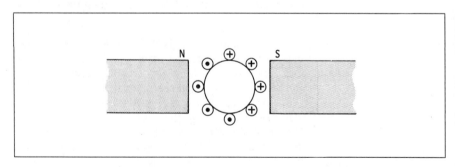

Generators

A magnetic field can produce a current in a conductor that is moving through that field.

If current flowing through a conductor produces a magnetic field, is it possible that the reverse is true—that a magnetic field can produce a current flow in a conductor? That is indeed the case. A dc generator operates according to this principle. An armature, similar to the one used in

UNDERSTANDING ELECTROMAGNETISM

a dc motor, is rotated in a fixed magnetic field. As the wires cut through the lines of flux, a current is generated in the wires. Brushes touching the split commutator ring lead the current to outside sources.

Apply a voltage source to the commutator connections of a small dc motor and the shaft turns. Turn the shaft of the same device and you can measure a voltage generated at the commutator connections. Small, permanent dc motors can operate as generators.

In a manner of speaking, a generator is simply a motor operated in reverse. You have seen that the interaction of current, flux lines, and force work for a motor in terms of a right-hand rule. It figures, then, that the same interactions apply to a generator, but as a left-hand rule. The left-hand rule for generators is illustrated in *Figure 8-15*.

**Figure 8-15.
Left-Hand Rule for Generators**

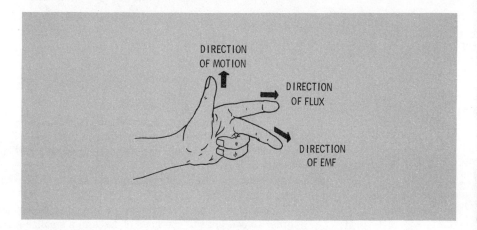

Meters

Meter movements work on the same magnetic principles as electric motors.

Meter movements—the kind normally used in voltmeters, ammeters, wattmeters, and so on—are actually small motors that can turn only about 150 degrees. Also, meters are spring loaded so that they return the movement to zero when the current source is removed.

As shown in *Figure 8-16*, the electromagnet in a meter movement is located between the poles of a permanent horseshoe magnet. When no current is flowing through the coil, a spring holds the pointer at the extreme left end of its travel. When current does flow, a magnetic field develops around the coil with the north pole on the left and the south on the right. The like poles repel and the coil rotates. How far the coil rotates depends on the strength of the electromagnet (the amount of current flowing through it) and the ability to overcome the tension applied by the spring.

**Figure 8-16.
Meter Principle**

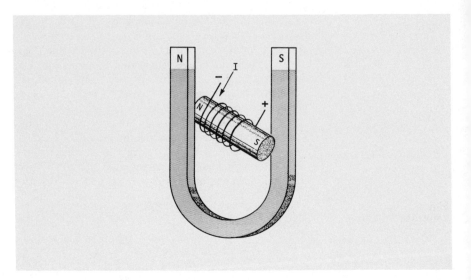

WHAT HAVE WE LEARNED?

1. An electromagnet is a temporary magnet with a magnetic field caused by a flow of current through a coil of wire.
2. An electromagnet retains its magnetic field only as long as current is flowing through its coil of wire.
3. Reversing the direction of current through an electromagnet reverses its magnetic polarity.
4. The coil of wire for an electromagnet is often wound around an iron core in order to concentrate the lines of force in a smaller area, thereby creating a stronger magnetic field.
5. The strength of an electromagnet is specified in terms of ampere-turns (the amount of current multiplied by the number of turns of wire in its coil).
6. Any magnetism that remains in the core of an electromagnet after the current has been turned off is called residual magnetism.
7. The discrepancy between the amount and direction of current flowing in an electromagnet and the strength and polarity of the core's magnetic field is called hysteresis.
8. The core of an electromagnet is said to be saturated when an increase in current no longer causes a corresponding increase in the strength of the magnetic field.
9. A solenoid is an electromagnet that has a movable core.
10. An electromagnet core that is shaped like a toroid confines the magnetic field to the core material.
11. A polarized electromagnet has a core that is permanently magnetized. The magnetic field from the coil either aids or opposes that of the core.
12. Electric motors operate on the principle that two magnetic fields interact to produce mechanical force.
13. The movable coil in a motor is called the armature.

UNDERSTANDING ELECTROMAGNETISM

14. A commutator makes up the electrical connection between the armature in a motor and the external source of electrical power.
15. Electric generators work on the principle that a magnetic field produces a current in a conductor moving through the field.
16. Meter movements operate on the same principle as electric motors, except that the meter movement is not permitted to turn through a full circle.

KEY WORDS

Ampere-turns
Armature
Commutator
Electromagnet
Generator
Hysteresis

Hysteresis loss
Motor
Residual magnetism
Saturation
Solenoid
Toroid

Quiz for Chapter 8

1. Which one of the following principles best describes the basic operation of an electromagnet?
   a. Like magnetic poles repel, while unlike poles attract.
   b. Current is induced in a conductor that is moving through a magnetic field.
   c. Current flowing through a conductor generates a magnetic field.
   d. Like magnetic poles attract, while unlike poles repel.

2. Electron current is flowing through a conductor in a direction you might describe as straight into this page. What is the direction of the resulting magnetic field?
   a. Clockwise.
   b. Counterclockwise.
   c. Into this page.
   d. Out of this page.
   e. There is not enough information to answer this question.

3. A certain electromagnet has 1000 turns of wire. How much current must flow through the coil in order to generate an mmf of 100 ampere-turns?
   a. 0.1 A.
   b. 1 A.
   c. 10 A.
   d. 100 A.

4. The core of an electromagnet is said to be in saturation when:
   a. the core is permitted to move freely within the coil.
   b. an increase in current through the coil causes no further increase in the mmf.
   c. changing the polarity of the current in the coil changes the polarity of the magnetic field around the core.
   d. frequent changes in the direction of current through the coil causes heating effects in the core material.

5. Which one of the following phrases best describes residual magnetism?
   a. Magnetism that remains in the core of an electromagnet after the current through the coil is turned off.
   b. The force of attraction between the poles of a permanent and temporary magnetic field.
   c. The circular magnetic field that surrounds a conductor carrying a current.
   d. The amount of magnetic flux that is not confined to the core material of an electromagnet.

UNDERSTANDING ELECTROMAGNETISM

6. Which one of the following statements best describes magnetic hysteresis?
   a. Hysteresis is most apparent when a steady amount of current is flowing through the coil of an electromagnet.
   b. Hysteresis is the amount of magnetic flux that is not confined to the core material of an electromagnet.
   c. Hysteresis is most apparent when no current is flowing through the coil of an electromagnet.
   d. Hysteresis is most apparent when the core material is saturated.
   e. Hysteresis is most apparent when the current through the coil of an electromagnet changes frequently.

7. The component that forms the electrical connection between the rotating coil of wire in a motor and the external source of electrical energy is called the:
   a. armature.
   b. rotor.
   c. battery.
   d. commutator.

8. The rotating coil of wire in an electric motor is called the:
   a. armature.
   b. brush.
   c. loop.
   d. commutator.

9. Which one of the following principles best describes the basic operation of a generator?
   a. Like magnetic poles repel; unlike poles attract.
   b. Current is induced in a conductor that is moving through a magnetic field.
   c. Current flowing through a conductor generates a magnetic field.
   d. Like magnetic poles attract; unlike poles repel.

Understanding Alternating Current

ABOUT THIS CHAPTER

When you have finished this chapter you will understand alternating current. You will know how alternating currents are generated and you will be able to recognize alternating and pulse waveforms.

Dc CURRENT FLOW

You are already familiar with the most important features of direct current (dc) and direct-current circuits. Voltage cells, batteries, and dc generators are the most common sources of dc electrical energy. Current in a dc circuit always flows in one direction only.

Figure 9-1 compares two simple dc circuits. The only difference between the two is that the batteries are connected in such a way that the electron current flows in different directions. Electron current flow is always from the negative terminal of a voltage source, through the circuit, and back to the positive terminal of the source. The battery in *Figure 9-1a* is connected in such a way that electron current flows through the circuit in clockwise direction. The battery in *Figure 9-1b* is connected so that electron flow is in a counterclockwise direction.

**Figure 9-1.
Electron Flow in Dc Circuits**

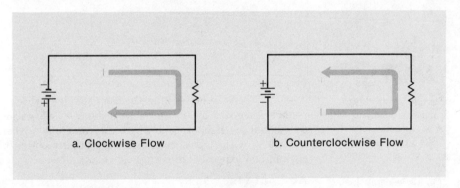

a. Clockwise Flow b. Counterclockwise Flow

AC CURRENT FLOW

Current in an ac circuit reverses direction periodically.

Now imagine that you are able to switch the polarity of the voltage source very rapidly. Part of the time current flows through the circuit in a clockwise direction, and part of the time it flows in a

9
UNDERSTANDING ALTERNATING CURRENT

counterclockwise direction. This is the basic nature of alternating current (abbreviated ac). In an ac circuit, the direction of current flow is reversed periodically.

In Common Use

Alternating current is more commonly used today. It is generated in great quantities in power plants. The electricity used in your home comes from such a power plant. The total current capacity of a power plant can reach several thousand amperes at 4 kV (4000 V).

Because the electron flow in ac circuits reverses direction regularly and rapidly, the value of an ac voltage can be easily increased or decreased by passing it through a transformer. For example, when it is necessary to transmit electricity from a power plant to a city many miles, as shown in *Figure 9-2*, the voltage from the power plant is stepped up to about 70 kV and sent through relatively small-diameter transmission wires. (To use a lower voltage level would require larger, heavier, and more expensive transmission wires because it would require larger amounts of current to send the same amount of electrical power.)

**Figure 9-2.
Transmitting Alternating Current**

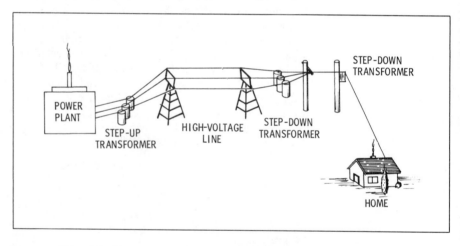

Electric power utility companies supply us with ac power because it is easier to change the level of ac voltage as required for efficient transmission from the generating station to the home.

When the high-voltage transmission lines reach the city, the voltage is reduced by transformers at a power sub-station. It is then distributed to neighborhoods where the voltage is further reduced to 240 and 120 V by the transformers we see on utility poles. It is far more difficult and expensive to change dc voltage levels.

Ac Sources

Most ac power is derived from ac generators. These work according to the principles of electromagnetism described in Chapter 8. Large amounts of mechanical energy are required to generate the tremendous amounts of ac electrical power required by modern civilization. Some ac generators obtain the necessary mechanical energy from flowing

UNDERSTANDING ALTERNATING CURRENT

water—water that is stored behind large dams. Other generators use steam energy that is derived from the heat of burning coal or a controlled atomic reaction.

Another source of ac, usually for specialized purposes, is the "oscillator." The current output from an oscillator is usually quite small—in the milliampere range. As a result, oscillators are used primarily as signal sources in electronic equipment.

WAVEFORMS

Ac and dc voltage and current can be plotted on a graph, called a waveform, showing the change in levels with the passage of time.

"Waveforms" are graphs showing how currents and voltages change over a period of time. The value of the voltage or current ("amplitude") is usually represented on the vertical plane, while time is represented on the horizontal plane.

Dc Waveforms

Figure 9-3 shows a voltage that rises from 0 V in a short period of time. Once the voltage reaches its maximum level, the graph indicates that the voltage remains fixed at that level indefinitely. This is a dc waveform because the voltage does not change polarity. The waveform is usually a straight line.

**Figure 9-3.
Dc Waveform**

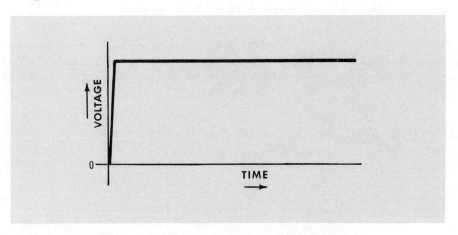

Figure 9-4 shows two other common kinds of dc waveforms. These are pulsating dc waveforms. The square-wave in *Figure 9-4a* shows the voltage resting at zero for a period of time and then switching to a maximum level where it remains for a time. It is then switched to zero to begin the cycle all over again. You would see this waveform in a dc circuit that is being switched on and off at a regular rate.

**Figure 9-4.
Pulsating Dc Voltage
Waveforms**

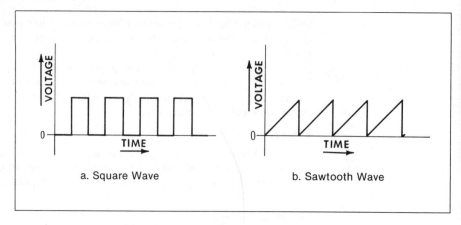

a. Square Wave   b. Sawtooth Wave

The sawtooth waveform in *Figure 9-4b* indicates a slightly different kind of change in voltage levels. The voltage increases gradually from 0 V. When it reaches a peak level, however, it drops immediately to 0 V.

The waveforms in *Figures 9-3* and *9-4* are all dc waveforms because the voltage always remains on the positive (upper) side of the voltage axis on the graph.

Ac Waveforms

The sine wave is the most common kind of ac waveform.

The most common ac waveform is the sine wave shown in *Figure 9-5*. In fact, the sine wave is so commonly used that when we think of alternating current, we automatically think of sine waves. You can see that it is an alternating-current waveform because it is sometimes positive and sometimes negative.

**Figure 9-5.
Sine Wave**

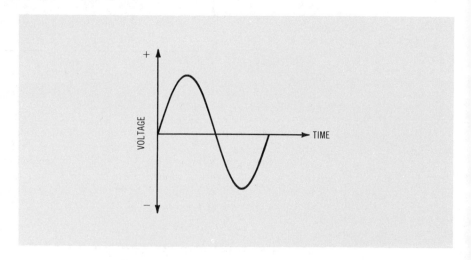

UNDERSTANDING ALTERNATING CURRENT

Generating Sine Waves

A loop of wire rotating in a magnetic field generates voltage and current that have a sine waveform.

A sine wave is the most common ac waveform and the simplest to create with electrical power generators. You can visualize the way the sine wave is generated by looking at *Figure 9-6*. As the coil cuts the lines of flux between the poles of the permanent magnet, a voltage is produced that will cause current to flow if the coil is connected to a complete circuit. The coil rotates in the indicated direction at a constant speed.

**Figure 9-6.
Coil Rotating Between Poles of Magnet**

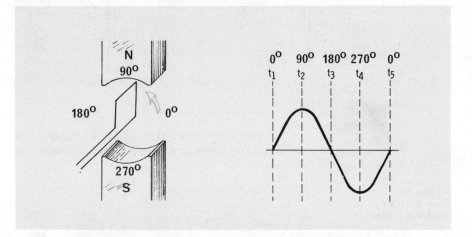

As the coil moves from 0° toward 90°, it cuts the flux lines at a more direct angle and the voltage increases. At 90°, it is at a right angle to the lines of flux, so it cuts the maximum number of lines per second. Therefore the voltage is maximum.

At 180°, the coil is parallel to the lines of flux and, therefore, is cutting through none of them. Thus, the voltage generated is zero. Beyond 180°, the coil cuts the flux lines in the opposite direction, so the generated voltage has the opposite polarity—negative in this case.

Another way to visualize the generation of a sine wave is to imagine a straight line that rotates something like the hand of a clock. This line can be called a voltage vector. As shown in *Figure 9-7*, however, a voltage vector rotates in the counterclockwise direction. The distance measured from the end of the voltage vector to the base line at any time during the rotation of the vector represents the exact value of voltage at that instant. You can see that the value of the voltage is zero at 0° and 180°. At 90°, the value of the voltage is maximum positive and, at 270°, it is maximum negative. One complete rotation of the vector represents one complete cycle of the sine wave. The time axis for a sine wave most often expresses the passage of time in terms of degrees of rotation of this voltage vector.

9 UNDERSTANDING ALTERNATING CURRENT

Figure 9-7.
Voltage Vector

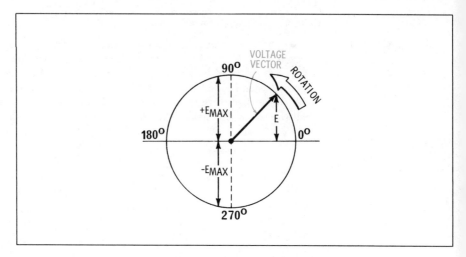

The simple sine wave is the basic building block for all other ac waveforms. As shown in *Figure 9-8*, even a square wave is a combination of sine waves.

Figure 9-8.
Sine Waves Are Building Blocks

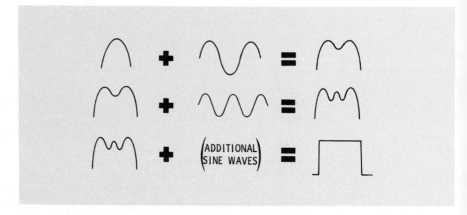

Measuring Sine Waves

Dc voltage sources provide single, steady voltage levels. Sine-wave voltage sources provide a voltage that is constantly changing. In *Figure 9-9*, you can see that the sine wave's voltage rises to a positive peak, drops gradually through zero to a negative peak, then rises gradually through zero to the positive peak.

UNDERSTANDING ALTERNATING CURRENT

Figure 9-9.
Sine Wave Peak Values

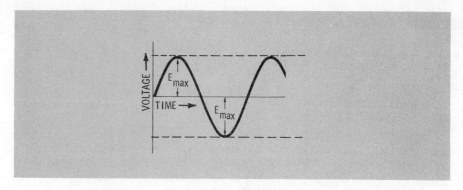

Notice that the voltage reaches a certain maximum level two times in every complete cycle—once positive, once negative. The voltage levels for these peaks have the same value but different polarities. The mathematical symbol for the maximum, or peak, voltage of a sine wave is E_{max}.

Rms Value

A more practical way to specify a sine-wave voltage level is by means of the "rms" voltage level (rms stands for root-mean-square). The rms value is the actual "working value" of a voltage or current, and it is equivalent to the dc value required for doing the same amount of work.

As shown in *Figure 9-10*, the rms value of a sine wave is 70.7% of its maximum value. The rms value is the one most commonly cited for ac voltage and current ratings. Power outlets in your home are said to supply 120 V. That is actually an rms value. The peak value is 1.41 times that amount, or about 170 V.

The actual working voltage of a sine wave is the root-mean-square (rms) value, which is equal to 0.707 times the peak value. You can assume that the voltage rating for a sine waveform is this rms value.

Figure 9-10.
RMS Value

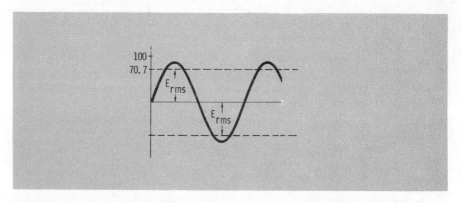

The mathematical symbol for rms voltage is E_{rms}. Unless otherwise specified, you can safely assume that a given ac voltage or current rating states rms value. You will find the following equations helpful for converting peak voltages to rms voltages and vice versa:

$$E_{rms} = 0.707 \times E_{max}$$
$$E_{max} = 1.41 \times E_{rms}$$

where,

E_{max} is a maximum, or peak, voltage reading,
E_{rms} is an rms, or working, voltage reading.

Like dc voltage, the unit of measurement for ac voltage is the volt. In order to make a clear distinction between the two units, dc voltages are labeled V dc while ac rms voltages are labeled V ac.

By way of an example, suppose you want to determine the peak value of a 240-V ac (rms) sine wave. The appropriate procedure is:

$$E_{max} = 1.41 \times E_{rms}$$
$$E_{max} = 1.41 \times 240$$
$$E_{max} = 274 \text{ V}$$

The current flowing in a circuit that is operated from an ac sine-wave voltage has the same waveform. A symbol for peak current is I_{max} and for rms current is I_{rms}. The relationships previously described for sine-wave voltages apply equally well to sine-wave currents.

$$I_{rms} = 0.707 \times I_{max}$$
$$I_{max} = 1.41 \times I_{rms}$$

where,

I_{max} is a maximum, or peak, current reading,
I_{rms} is an rms, or working, current reading.

Sine-Wave Cycle

Sine waves are also specified according to the amount of time required to complete one full cycle and according to the number of cycles that occur each second.

You have seen how an ac voltage first increases in value to a maximum, then decreases through zero to a maximum of the opposite polarity, and then rises back to zero. This sequence is known as one complete cycle. The waveform in *Figure 9-11* shows three complete cycles of a sine wave.

A cycle is one complete change from zero to the positive peak value, back through zero to the negative peak value, and back to zero. This represents the rotation of a voltage vector around 360° of a circle.

UNDERSTANDING ALTERNATING CURRENT

**Figure 9-11.
Three Cycles of a Sine Wave**

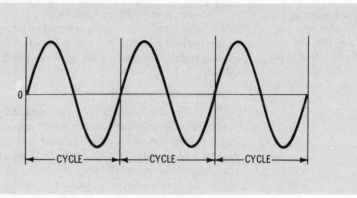

Sine-Wave Period and Frequency

The period of a waveform is the time required for completing one cycle.

The time required for running one complete cycle of a waveform is called the period of that waveform. The unit of measurement for the period of waveforms is the second, and the mathematical symbol is usually shown as T.

The frequency of a waveform is the number of complete cycles that occur each second.

The frequency of a waveform is the number of cycles completed each second. The ac electrical power supplied to your home completes 60 cycles each second. The unit of measurement for frequency is hertz (Hz), and the mathematical symbol is f.

The ac power available in the home can thus be completely specified in terms of the rms voltage and the frequency. Those values happen to be 120 V ac, 60 Hz.

There is a simple relationship between the period of a waveform and its frequency:

$$f = \frac{1}{T}$$

$$T = \frac{1}{f}$$

where,

f is the frequency in Hz,
T is the period in sec.

What is the period of a common 60-Hz waveform?

$$T = \frac{1}{f}$$

$$T = \frac{1}{60}$$

$$T = 0.0167 \text{ sec}$$

1 ms = 1/1000 sec
1 μs = 1/1,000,000 sec
1 kHz = 1000 Hz
1 MHz = 1,000,000 Hz

In most practical applications, the periods of waveforms are rather small values and frequencies are rather large values. It is not uncommon, to see periods listed in units of milliseconds (ms) or microseconds (μs). The corresponding frequencies are given in kilohertz (kHz) and megahertz (MHz).

Measuring Pulse Waveforms

The sine wave is the most commonly encountered waveform, but it is not the only one that has important applications. Square-wave pulses, for instance, are commonly used in radar and digital computers. As shown in *Figure 9-12*, a radar system sends out a brief pulse of radio energy traveling at the speed of light. When the square-wave pulse strikes an object, such as an airplane, part of the pulse energy is reflected back to the radar. The radar keeps track of the time it takes the pulse to travel to the object and return. This time interval can be used for calculating the exact distance to the object.

**Figure 9-12.
Radar Uses Square-Wave Pulses**

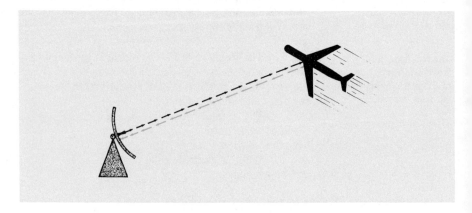

Digital computers use square waveforms for performing lengthy and complex counting and logical operations. A carefully controlled oscillator generates the square waves that time all of the computer's operations. The oscillator in personal computers operates at a frequency on the order of 4 MHz. Larger, high-performance computers operate at frequencies of 10 MHz and 20 MHz.

Square-wave pulses are measured according to their maximum voltages. Like sine waves, square waves can also be specified in terms of period and frequency.

Pulse waveforms of any shape are usually described in terms of four parts: baseline, leading edge, peak, and trailing edge. These are shown applied to the waveforms in *Figure 9-13*.

The four main parts of a pulse waveform are the baseline, the leading edge, the peak, and the trailing edge.

**Figure 9-13.
Four Parts of Pulse Waveforms**

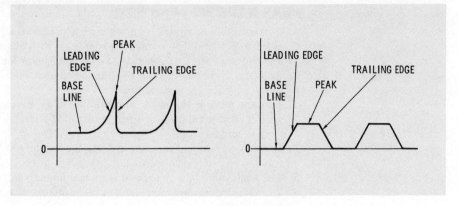

Rise and Decay Time

It is often necessary to know how long it takes a pulse to rise from the baseline to its peak. This interval is called the "rise time" of the waveform. The time required to drop from the peak to the base line is called the "decay time" of the waveform.

Figure 9-14 shows that the rise time of a waveform is the time required to change from 10% of the peak to 90% of the peak. The decay time is the interval required for the waveform to fall from 90% of its peak value to the 10% level.

**Figure 9-14.
Rise and Decay Times of Pulse Waveforms**

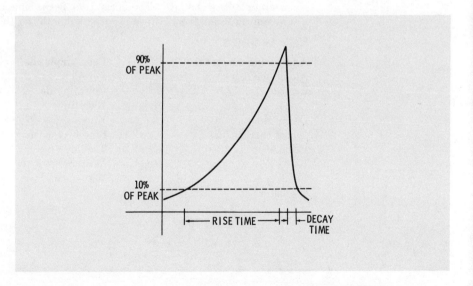

9 UNDERSTANDING ALTERNATING CURRENT

WHAT HAVE WE LEARNED?

1. Current in an ac circuit reverses direction periodically.
2. Electric power utility companies supply us with ac power because it is easier and less expensive to generate and transmit over long distances.
3. Ac voltage and current can be graphed as a waveform, showing the change in levels with the passage of time.
4. The sine wave is the most common kind of ac waveform.
5. A loop of wire rotating in a magnetic field generates sine waveforms of voltage and current.
6. One full, 360-degree rotation of a voltage vector produces a complete sine wave.
7. The peak value of a sine wave is its maximum positive or negative level.
8. The actual working voltage and current of sine waves is the rms (root-mean-square) value.
9. The rms value of a sine wave is equal to 0.707 times the peak value.
10. A cycle is one complete change from zero to the positive peak value, back through zero to the negative peak value, and back to zero.
11. The period of a waveform is the time required to complete one cycle.
12. The frequency of a waveform is the number of complete cycles that occur each second. The unit of measurement for frequency is the hertz (Hz).
13. The four main parts of a pulse waveform are the baseline, leading edge, peak, and trailing edge.
14. The rise time of a waveform is the time required to change from 10% of the peak to 90% of the peak. The decay time is the time required for the waveform to fall from 90% of its peak value to the 10% level.

KEY WORDS

Alternating current (ac)
Amplitude
Baseline
Cycle
Decay time
Direct current (dc)
Frequency
Hertz (Hz)
Leading edge
Oscillator

Peak amplitude
Period
Pulsating dc
Rise time
Rms
Sine waveform
Trailing edge
Vector
Waveform

Quiz for Chapter 9

1. Which one of the following statements most accurately reflects the nature of ac power?
   a. The current in an ac circuit flows in one direction at a steady level.
   b. The current in an ac circuit flows in one direction, but changes value periodically.
   c. The current in an ac circuit changes direction periodically.
   d. The current in an ac circuit flows in two directions at the same time.

2. Most ac power is obtained from:
   a. batteries.
   c. generators.
   d. oscillators.
   e. lightning.

3. Waveforms are graphs that show:
   a. amplitude on the vertical axis and frequency on the horizontal axis.
   b. frequency on the vertical axis and amplitude on the horizontal axis.
   c. amplitude on the vertical axis and the passage of time on the horizontal axis.
   d. the passage of time on the vertical axis and amplitude on the horizontal axis.

4. Which one of the following phrases most accurately describes the nature of a pulsating dc voltage?
   a. The voltage remains at one polarity and at a steady level.
   b. The voltage remains at one polarity but changes value periodically.
   c. The voltage reverses polarity periodically.
   d. The voltage causes current to flow in two directions at the same time.

5. The peak value of a sine waveform is:
   a. the amplitude at 90% of the maximum value.
   b. the amplitude at 70.7% of the maximum value.
   c. the amplitude at 1.41 times the maximum value.
   d. the maximum positive or negative value.

6. The rms value of a sine waveform is:
   a. the time required to complete one full cycle.
   b. the number of cycles completed in one second.
   c. equal to 0.707 times the peak amplitude.
   d. equal to 1.41 times the peak amplitude.
   e. the average amplitude.

7. The period of a waveform is:
   a. the time required to complete one full cycle.
   b. the number of cycles completed in one second.
   c. equal to 0.707 times the peak amplitude.
   d. the time required for the voltage to rise from 10% to 90% of the peak amplitude.

8. Which one of the following statements most accurately applies to the frequency of a waveform?
   a. Unless specified otherwise, the frequency of a waveform is assumed to be 60 Hz.
   b. The actual frequency of a sine waveform is 1.41 times the working frequency.
   c. The higher the amplitude of a waveform, the higher its frequency.
   d. The longer the period of a waveform, the lower its frequency.

9. The rise time of a pulse waveform is the time required for the voltage to rise:
   a. from zero to its peak value.
   b. from zero to its rms value.
   c. from 10% of the peak value to 70.7% of the peak value.
   d. from 10% of the peak value to 90% of the peak value.

10

Working with Resistive Ac Circuits

ABOUT THIS CHAPTER

In this chapter, you will learn how to draw the schematic of a basic ac circuit. You will learn when and how Ohm's law and the power formulas can be used in ac circuits. You will learn how to simplify resistance combinations in ac circuits to find the equivalent resistance. You also will learn to tell when voltage and current are in phase.

A BASIC AC CIRCUIT

The basic ac circuit is very similar to the basic dc circuit. The only difference is that an ac generator is used as a voltage source in ac circuits instead of a battery or dc generator. In any ac circuit, the voltage shown at the generator is assumed to be the rms voltage level. The frequency of the ac source is specified in hertz (Hz) and the circuit resistance in ohms. See the example in *Figure 10-1*.

**Figure 10-1.
Series Ac Circuit**

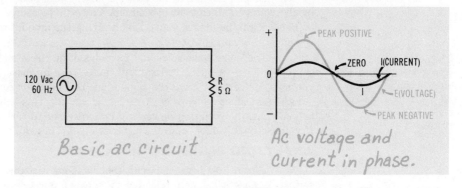

Basic ac circuit.

Ac voltage and current in phase.

Unless stated otherwise, ac voltage and current levels are expressed in rms units.

Ohm's law can be applied to ac circuits so long as the circuit is purely resistive. (Circuits that are not resistive are described in the remaining chapters.) As stated previously, the rms voltage is the working voltage level of ac electricity. The same is true for rms current—it is the working current level for ac circuits. In a basic ac circuit, all calculations are made with rms voltage and current levels, unless otherwise specified.

10 WORKING WITH RESISTIVE AC CIRCUITS

In the circuit in *Figure 10-1*, you should be able to use Ohm's law to determine the amount of current flowing through it:

$$I = \frac{E}{R}$$

$$I = \frac{120}{5}$$

$$I = 20 \text{ A}$$

Applying 120 V rms to a resistance of 5 Ω thus forces 20 A rms to flow.

The voltage and current in the circuit in *Figure 10-1* are both sine waveforms, but they have different values and units of measurement. When the voltage and current waveforms rise (increase) and fall (decrease) at exactly the same time, they are said to be "in phase." The waveforms in *Figure 10-1* are in phase. When a circuit has only pure resistance, the voltages and currents are in phase with one another. Ohm's law applies directly in ac circuits as in dc circuits so long as you use the same kinds of values (usually rms) for the voltage and current.

When the voltage and current waveforms do not cross zero and reach their peaks at the same instant, they are said to be "out of phase." Ohm's law does not always apply directly to circuits in which the voltage and current are out of phase. Later chapters explain out-of-phase voltage and current waveforms.

As you learned in the study of dc circuits, power is the work done by the voltage and current in a circuit. Power is measured in units of watts (W), and it can be determined by multiplying the amount of voltage times the amount of current ($P = E \times I$).

The power principles and formulas of dc circuits also apply in ac circuits so long as the ac voltage and current are in phase. Furthermore, you must be certain to use the same kinds of values for current and voltage—both rms values, or both peak values. If you use rms values for voltage and current, the resulting power will be expressed in terms of rms watts. If you use peak values, power will be expressed in peak watts.

ANALYZING SERIES AC CIRCUITS

Figure 10-2 is an ac circuit that has four resistors connected in series. Because it has only resistors in it, it is a purely resistive circuit. Currents and voltages in resistive ac circuits are in phase, so you can make the necessary calculations according to the standard Ohm's law and power equations.

Ac current and voltage are in phase when their levels rise (increase) and fall (decrease) together. All Ohm's law equations apply to circuits where current and voltage are in phase.

Ac current and voltage are out of phase when they cross zero and reach peak values at different times.

The power formulas and principles for power in dc circuits apply to ac circuits so long as the current and voltage are in phase.

WORKING WITH RESISTIVE AC CIRCUITS 10

Figure 10-2.
Series Ac Circuit Analysis

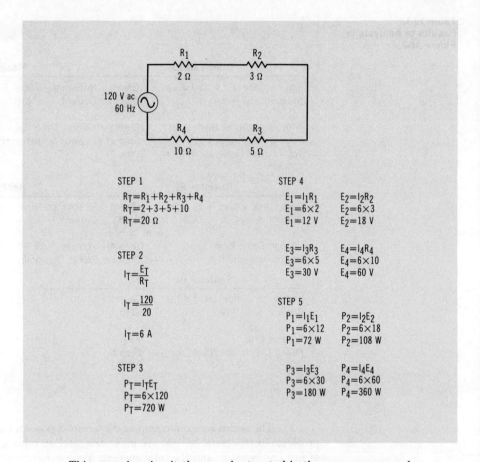

STEP 1
$R_T = R_1 + R_2 + R_3 + R_4$
$R_T = 2 + 3 + 5 + 10$
$R_T = 20\ \Omega$

STEP 2
$I_T = \dfrac{E_T}{R_T}$

$I_T = \dfrac{120}{20}$

$I_T = 6\ A$

STEP 3
$P_T = I_T E_T$
$P_T = 6 \times 120$
$P_T = 720\ W$

STEP 4
$E_1 = I_1 R_1$ $E_2 = I_2 R_2$
$E_1 = 6 \times 2$ $E_2 = 6 \times 3$
$E_1 = 12\ V$ $E_2 = 18\ V$

$E_3 = I_3 R_3$ $E_4 = I_4 R_4$
$E_3 = 6 \times 5$ $E_4 = 6 \times 10$
$E_3 = 30\ V$ $E_4 = 60\ V$

STEP 5
$P_1 = I_1 E_1$ $P_2 = I_2 E_2$
$P_1 = 6 \times 12$ $P_2 = 6 \times 18$
$P_1 = 72\ W$ $P_2 = 108\ W$

$P_3 = I_3 E_3$ $P_4 = I_4 E_4$
$P_3 = 6 \times 30$ $P_4 = 6 \times 60$
$P_3 = 180\ W$ $P_4 = 360\ W$

This ac series circuit thus can be treated in the same way as a dc series circuit. Step 1 in the analysis in *Figure 10-2* shows that the total resistance of the circuit is 20 Ω. In Step 2 you use Ohm's law to calculate the total current. In Step 3 you use the basic power formula to determine the total power dissipation.

In Step 4 you use Ohm's law to calculate the voltage drop across each resistor, based on the value for total current you found in Step 2. This step takes advantage of the fact that the current through each resistor in a series circuit is equal to the total circuit current. In Step 5 you determine the rms power dissipation for each resistor, based on the current and voltage for each one of them.

Table 10-1 summarizes the results of this analysis.

10 WORKING WITH RESISTIVE AC CIRCUITS

Table 10-1.
Results of Analysis in Figure 10-2.

Total Values		Resistor R_1	
Total voltage: $E_T = 120$ V ac	(Given)	Resistance Value: $R_1 = 2\ \Omega$	(Given)
Frequency: $f = 60$ Hz	(Given)	Current: $I_1 = 6$ A	(Step 2; series circuit)
Total Resistance: $R_T = 20\ \Omega$	(Step 1)	Voltage Drop: $E_1 = 12$ V	(Step 4)
Total Current: $I_T = 6$ A	(Step 2)	Power Dissipation: $P_1 = 726$ W	(Step 5)
Total Power Dissipation: $P_T = 720$ W	(Step 3)		

Resistor R_2		Resistor R_3	
Resistance Value: $R_2 = 3\ \Omega$	(Given)	Resistance $R_3 = 5\ \Omega$	(Given)
Current: $I_2 = 6$ A	(Step 2; series circuit)	Current: $I_3 = 6$ A	(Step 2; series circuit)
Voltage Drop: $E_2 = 18$ V	(Step 4)	Voltage Drop: $E_3 = 30$ V	(Step 4)
Power Dissipation: $P_2 = 108$ W	(Step 5)	Power Dissipation: $P_3 = 180$ W	(Step 5)

Resistor R_4	
Resistance Value: $R_4 = 10\ \Omega$	(Given)
Current: $I_4 = 6$ A	(Step 2; series circuit)
Voltage Drop: $E_4 = 60$ V	(Step 4)
Power Dissipation: $P_4 = 360$ W	(Step 5)

The series resistor circuit in *Figure 10-3* poses a different kind of problem. In this instance, you are given the total ac voltage level, the value of just one resistor, and the total current. Your objective is to calculate the value of the unknown resistance, R_2.

**Figure 10-3.
Determining Resistor Value in Series Circuits**

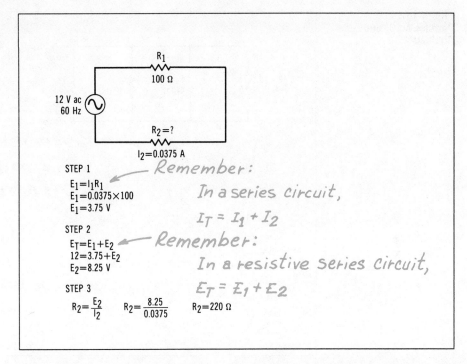

The general rule for approaching problems of this type is to begin wherever you can. In this case, you can calculate the voltage drop across resistor R_1 based on the given value of its resistance and the given amount of current flowing through this series circuit. The calculation is shown in Step 1. Knowing the total voltage and the voltage drop across resistor R_1, you now can determine the voltage across resistor R_2. As shown in Step 2, E_2 is 8.25 V.

At this point, you know the voltage drop and current for R_2. This is sufficient for applying Ohm's law to determine R_2 resistance, as shown in Step 3. The result is a value of 220 Ω.

ANALYZING PARALLEL AC CIRCUITS

Figure 10-4 shows a parallel circuit operated from a 48-V, 400-Hz source. Because the circuit is purely resistive, the current and voltage are in phase with one another. So the circuit can be analyzed with the same Ohm's law and power formulas that apply to parallel dc circuits.

**Figure 10-4.
Parallel Ac Circuit Analysis**

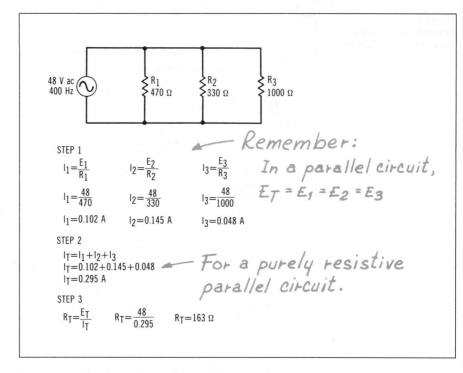

Use Step 1 in the analysis to determine the ac current for each resistor. This step takes advantage of the fact that the voltage is the same across each branch of a parallel circuit.

Use Step 2 to add up the currents through the individual resistors to determine the total circuit current. This step is valid only for purely resistive parallel circuits.

In the final step, you use Ohm's law to calculate the total resistance on the basis of the total voltage and current. *Table 10-2* summarizes the results of a complete analysis of the circuit.

**Table 10-2.
Results of Analysis in Figure 10-4**

Total Values		Resistor R_1	
Total Voltage: E_T = 48 V	(Given)	Value of R_1 = 470 Ω	(Given)
Total Current: I_T = 0.295 A	(Step 2)	Voltage for R_1 = 48 V	(Given, parallel circuit)
Total Resistance: R_T = 163 Ω	(Step 3)	Current for R_1 = 0.102 A	(Step 1)

WORKING WITH RESISTIVE AC CIRCUITS

Table 10-2 cont.

Resistor R_2		Resistor R_3	
Value of R_2 = 330 Ω	(Given)	Value of R_3 = 1000 Ω	(Given)
Voltage for R_2 = 48 V	(Given, parallel circuit)	Voltage for R_3 = 48 V	(Given, parallel circuit)
Current for R_2 = 0.145 A	(Step 1)	Current for R_3 = 0.048 A	(Step 1)

The circuit in *Figure 10-5* requires you to handle the analysis of a parallel circuit in a different fashion. You are given the total voltage and the wattage rating of two lamps. The problem is to determine the total amount of current for the circuit.

Figure 10-5. Total Current in Lamp Circuit

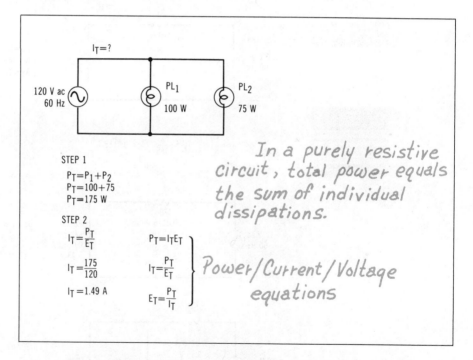

In a purely resistive circuit, total power equals the sum of individual dissipations.

Power/Current/Voltage equations

The calculations in Step 1 take advantage of the fact that the total power dissipated in a purely resistive circuit is equal to the sum of the individual power figures. The total power dissipation in this instance is 175 W (rms).

Having determined the total power dissipation, you can apply one of the power formulas to calculate the total current. The appropriate formula in this case is $I_T = P_T/E_T$. Step 2 shows how this formula applies.

10 WORKING WITH RESISTIVE AC CIRCUITS

ANALYZING COMBINATION AC CIRCUITS

As long as an ac circuit contains nothing but resistive components, the procedures for analyzing the circuit are identical to the procedures already described for dc circuits. *Figure 10-6* shows an analysis of a combination circuit (series or parallel).

Figure 10-6.
Combination Ac Circuit Analysis

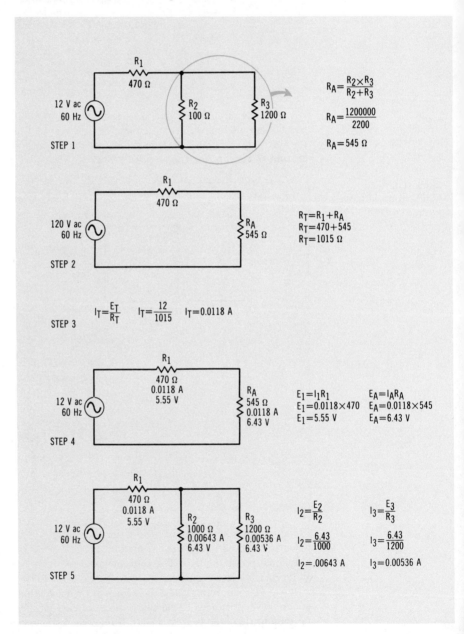

WORKING WITH RESISTIVE AC CIRCUITS 10

Use the product-over-sum rule in Step 1 to determine the equivalent resistance for the parallel combination of resistors R_2 and R_3. Step 2 indicates that the equivalent resistance, R_A, is in series with R_1. The total resistance of the circuit is the sum of the two resistances. In Step 3, use the values of total voltage and resistance to determine the total current for the circuit.

We begin putting the circuit together again in Step 4. Because R_1 and R_A are connected in series, we know that their currents are equal to the total circuit current. We use Ohm's law to determine the voltage drops across these two resistances.

We complete the analysis in Step 5 by calculating the currents through resistors R_2 and R_3. See the summary of this analysis in *Table 10-3*.

**Table 10-3.
Results of Analysis in Figure 10-6**

Total Values		Resistor R_1	
Total Voltage: E_T = 12 V	(Given)	Value of R_1 = 470 Ω	(Given)
Total Current: I_T = 0.0118 A	(Step 3)	Voltage for R_1 = 5.55 V	(Step 4)
Total Resistance: R_T = 1015 Ω	(Step 2)	Current for R_1 = 0.0118 A	(Step 3)
Resistor R_2		**Resistor R_3**	
Value of R_2 = 1000 Ω	(Given)	Value of R_3 = 1200 Ω	(Given)
Voltage for R_2 = 6.43 V	(Step 4)	Voltage for R_3 = 6.43 V	(Step 4)
Current for R_2 = 0.00643 A	(Step 5)	Current for R_3 = 0.00536 A	(Step 5)

WHAT HAVE WE LEARNED?

1. Ac circuits are similar to dc circuits, but ac circuits use an ac source (such as a generator) in place of the dc source (such as a battery).
2. Unless stated otherwise, voltage, current, and power values are expressed in rms units.
3. Voltage, current, and power waveforms are in phase when they cross through zero and reach their positive and negative peaks at the same time.
4. Voltage, current, and power waveforms are out of phase when they cross through zero and reach their peaks at different times.
5. All Ohm's law and power equations apply when current and voltage are in phase.
6. A resistive circuit is an ac or dc circuit that contains only resistances.
7. Current and voltage are in phase in a purely resistive circuit.

10 WORKING WITH RESISTIVE AC CIRCUITS

Quiz for Chapter 10

1. Current and voltage are said to be in phase when:
   a. they have the same amplitude and frequency.
   b. their waveforms are both ac or both dc.
   c. their waveforms cross through zero and reach positive and negative peaks at the same time.

2. What is the total resistance of the circuit in *Figure 10-7*?
   a. 33.3 Ω.
   b. 50 Ω.
   c. 75 Ω.
   d. 100 Ω.
   e. 150 Ω.

3. What is the total ac current through the circuit in *Figure 10-7*?
   a. 0.060 A.
   b. 0.120 A.
   c. 0.180 A.
   d. 0.240 A.
   e. 0.541 A.

4. What is the total ac power dissipation of the circuit in *Figure 10-7*?
   a. 1.08 W.
   b. 2.16 W.
   c. 3.24 W.
   d. 4.32 W.
   e. 9.74 W.

Figure 10-7.
Circuit for Quiz Questions 2–4

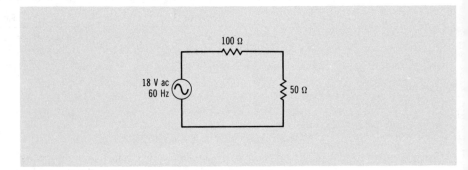

5. What is the power dissipation of the lamp in *Figure 10-8*?
   a. 2 W.
   b. 60 W.
   c. 100 W.
   d. 120 W.
   e. 200 W.
   f. 240 W.

6. What is the total ac power dissipation of the heater circuit in *Figure 10-9*?
   a. 333 W.
   b. 500 W.
   c. 750 W.
   d. 1000 W.
   e. 1500 W.

**Figure 10-8.
Circuit for Quiz
Question 5**

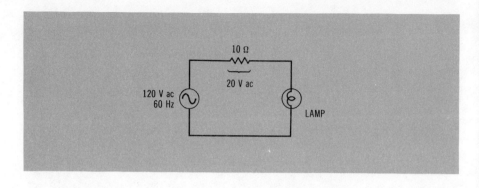

7. How much current flows through the 500-W heater in *Figure 10-9*?
   a. 2.76 A.
   b. 4.17 A.
   c. 6.25 A.
   d. 12.5 A.
   e. 83.3 A.

8. What is the resistance of the 1000-W heater in *Figure 10-9*?
   a. 0.18 Ω.
   b. 1.20 Ω.
   c. 9.60 Ω.
   d. 14.4 Ω.
   e. 19.2 Ω.

**Figure 10-9.
Circuit for Quiz
Questions 6–9.**

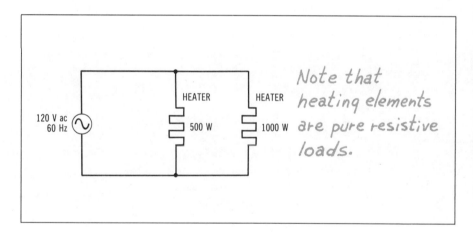

Note that heating elements are pure resistive loads.

9. What is the total ac current for the heater circuit in *Figure 10-9*?
   a. 2.76 A.
   b. 4.17 A.
   c. 6.25 A.
   d. 12.5 A.
   e. 83.3 A.

11

The Basics of Inductance

ABOUT THIS CHAPTER

Inductance is an important feature in many kinds of electrical circuits. Just as we insert a known amount of resistance into a circuit, there are also situations when it is important to insert a known amount of inductance. And just as some circuits include amounts of unwanted resistance that produces heating effects, some circuits include unwanted inductance that distorts waveforms and causes other difficulties.

This chapter describes the basic principles of inductance. You will learn about a few electrical devices that depend on inductance for their operation, and you will see how inductance in a circuit affects ac and dc voltage and current.

INDUCTANCE

When current begins flowing through a conductor, the current creates magnetic lines of force around the conductor. As this magnetic field grows in strength, its magnetic lines of force cut the conductor and generate a voltage that opposes the increasing current. Likewise, decreasing the current through a conductor causes the lines of force to collapse. As the magnetic field collapses, the lines of force cut the conductor and generate a voltage that bolsters the decreasing current.

Any change in current through a conductor thus generates a voltage in the conductor that tends to oppose the change. This self-induced voltage is called a "counter emf." The counter emf increases as the current changes more rapidly. In fact, the counter emf is proportional to the rate of change of current and always opposes that change.

Any conductor that carries a current is capable of showing the effects of counter emf. This is called the property of electrical "inductance." The unit of measurement for inductance is the henry (abbreviated H). A conductor is said to have an inductance of 1 H when current changing at the rate of 1 A per second produces a counter emf of 1 V.

A straight length of wire possesses very little inductance. A tightly wound coil of wire, on the other hand, can have a lot of inductance, especially if its magnetic fields are concentrated within a soft-iron core material. A coil of wire used for this purpose is called an "inductor." Because of this feature of its construction, an inductor is sometimes called a coil. (Older schematics and textbooks sometimes refer to an inductor as a choke.) *Figure 11-1* shows the schematic symbols for inductors.

A change in current through a conductor induces a counter emf proportional to the rate of change of current, with a polarity that opposes the change in current.
The unit of measurement for inductance is the henry (H). An inductance of 1 H produces 1 V of counter emf when the current is changing at the rate if 1 A per second. A practical inductor is made of a coil of wire.

11 THE BASICS OF INDUCTANCE

**Figure 11-1.
Schematic Symbols for Inductors**

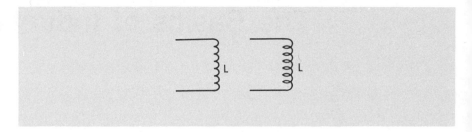

Figure 11-2a shows a changing current being applied to an inductor. The current first increases from 0 to 1 A in one second, then decreases from 1 A to 0 in the same period of time (a triangular waveform of current). *Figure 11-2b* shows the counter emf that the current waveform produces in a 1-H inductor. The figure shows the counter emf fixed at -1 V as long as the current is increasing. The counter emf is fixed at $+1$ V as long as the current is decreasing.

**Figure 11-2.
Changing Current Creates Counter Emf**

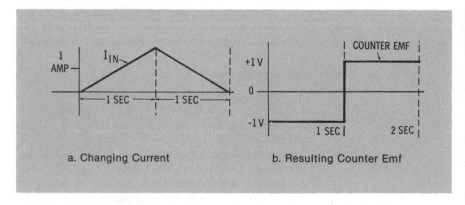

a. Changing Current   b. Resulting Counter Emf

A counter emf is generated in an inductance only when there is a change in the amount of current through the inductance.

Any change in current through an inductance creates a counter emf. The amount of counter emf depends on the rate of change of current. The polarity of the counter emf depends on the direction of change of current (increasing or decreasing). When there is no change in the amount of current, there can be no counter emf, no matter how much current is flowing.

The effect of counter emf in an inductor is quite dramatic when you apply a sine wave of current. As shown in *Figure 11-3*, a sine wave of current shows no change in level at its peaks (points 2 and 4), but a maximum rate of change where the sine wave crosses 0 (points 1 and 3). As a result, there is no counter emf induced when the current is at its peaks and the induced counter emf is at a maximum value where the current waveform crosses 0.

THE BASICS OF INDUCTANCE

**Figure 11-3.
Critical Points on
Current Sine Wave**

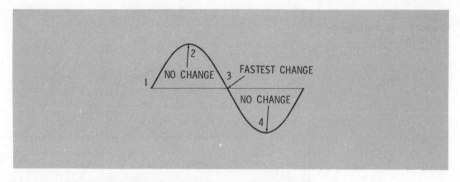

HOW INDUCTANCE AFFECTS AC CURRENT

If a sine wave of voltage is applied across a resistor, the current through the resistor also has a sine-wave form. At every instant through the cycle, the current can be determined by Ohm's law: I = E/R. As shown in *Figure 11-4*, the voltage and current sine waves are in phase.

**Figure 11-4.
Current and Voltage in
Phase Through a
Resistor**

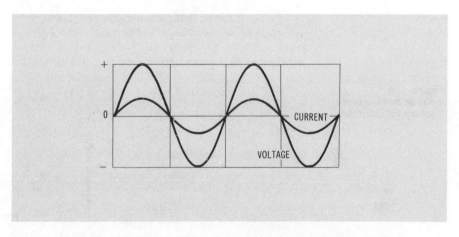

Sine waves of voltage and current for an inductor are out of phase.

Inductance opposes a change in current. The voltage value of a sine wave is always changing and, therefore, always trying to change the current through an inductance. This means that an inductance responds to a sine wave of voltage by retarding the change in current. The result is a current waveform that is delayed after the applied voltage waveform. Current and voltage are out of phase in an inductor.

The current through an inductor lags the applied voltage by 90 degrees.

Any sine waveform that is delayed with respect to another sine waveform is said to lag that waveform. As shown in *Figure 11-5*, the current through an inductor lags the voltage. The amount of lag is expressed in degrees. The current in an inductor reaches its peaks and crosses zero 90° later than the voltage waveform does. Therefore, current through an inductor lags the voltage by 90°.

11 THE BASICS OF INDUCTANCE

**Figure 11-5.
Current Lags Voltage by 90° in Inductor**

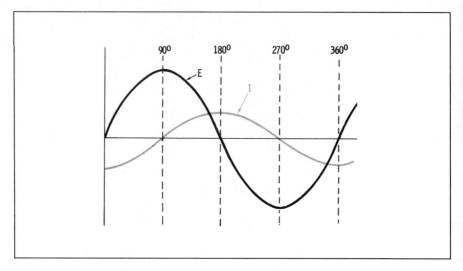

In a circuit that contains only resistance, ac current and voltage are in phase, so the voltage and current vectors always share the same position. *Figure 11-6* shows the voltage and current vectors for an inductance. Notice that the current vector is 90° behind the voltage vector.

**Figure 11-6.
Voltage and Current Vectors for Inductor**

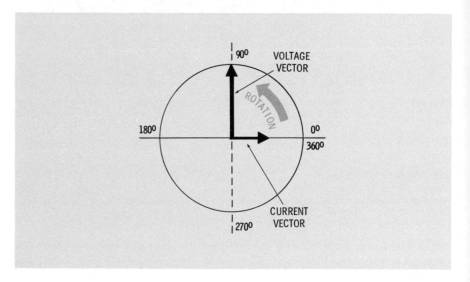

160   UNDERSTANDING ELECTRICITY AND ELECTRONICS CIRCUITS

THE BASICS OF INDUCTANCE

11

An inductance borrows energy from a circuit when current increases, but returns that same energy when current decreases. Unlike a resistance, an inductance does not consume electrical energy.

Inductance, unlike resistance, consumes no power. You'll recall that a resistor consumes power, in the form of heat, according to the power equation, $P = IE$. Inductance consumes no power and, therefore, dissipates no heat. However, when the current through an inductance is increasing, the inductance consumes energy that is necessary for building up the magnetic lines of force. When the current subsequently decreases, the collapsing lines of force return the built-up energy to the circuit. You can say that an inductance borrows energy from a circuit but always returns it.

VALUE OF INDUCTANCE

Inductance is proportional to the square of the number of turns in the coil.

Several factors determine the amount of inductance of an inductor, or coil. The inductance of a coil is proportional to the square of the number of turns of wire. *Figure 11-7* illustrates this principle. If a certain coil has twice as many turns of wire as another, it will have four times as much inductance. If it has three times as many turns, it will have nine times as much inductance, and so on.

**Figure 11-7.
More Turns Mean More Inductance**

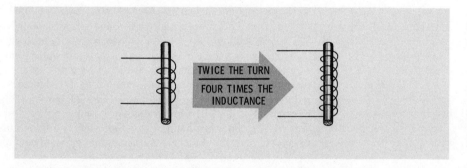

Inductance is proportional to the diameter of the coil of wire.

The diameter of the windings in an inductor also affects its inductance. As indicated in *Fig. 11-8*, the larger the diameter, the larger the amount of inductance.

**Figure 11-8.
Larger Coil Diameter Means More Inductance**

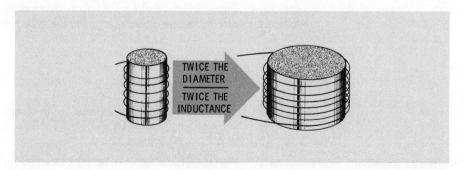

UNDERSTANDING ELECTRICITY AND ELECTRONICS CIRCUITS 161

11 THE BASICS OF INDUCTANCE

A coil that has an iron core has much more inductance than one having an air core.

As we saw earlier, placing an iron core in the center of a coil increases the coil's inductance because the iron sustains a much greater magnetic field than air. As we have seen, the inductance of a coil is related to the amount of magnetism it can produce.

Formulas are available for calculating the inductance of various types of coils. And there are tables to consult for simple, one-layer coils. Using these formulas and tables, you can design a coil that has a desired amount of inductance. You can also use the formulas and tables to determine the value of an inductor when you are able to count the number of turns, measure its diameter, and determine the properties of its core material.

INDUCTIVE REACTANCE

When current changes in an inductor, the property of inductance is responsible for producing a counter emf that opposes the change in current. In a dc circuit, this effect is present only when the source voltage is first applied and when it is first removed. The inductive effect lasts only as long as the dc current is rising to its maximum or is dropping to 0.

Opposition to ac current flow through an inductor is proportional to the frequency of the waveform.

In ac circuits, however, the current is constantly changing. Therefore, inductance is constantly working to oppose changes in current flow. The faster the current changes, the more opposition there will be. The rate of change of current in an ac circuit depends on the frequency of the waveform. The higher the frequency, the faster the current changes. So, an inductor offers far more opposition to current flow in a high-frequency circuit than it does in a low-frequency circuit.

On the other hand, a resistance offers a fixed amount of opposition to current flow, no matter what the circuit's frequency might be. A 100-Ω resistor, for example, offers 100 Ω resistance in a dc circuit and 100 Ω resistance in a 100-MHz circuit. An inductor also opposes current flow, but the amount of opposition is proportional to the frequency.

An inductor's opposition to current flow in an ac circuit is called inductive reactance.

The opposition to ac current flow through an inductor is called "inductive reactance." Like resistance, inductive reactance is measured in ohms. The mathematical symbol for inductive reactance is X_L, and the equation for determining the amount of inductive reactance is:

$$X_L = 2\pi f L$$

where,

X_L is the amount of inductive reactance in ohms,
π is the constant, pi, or 3.14,
f is the frequency in hertz,
L is the inductance in henrys.

THE BASICS OF INDUCTANCE

This equation shows that inductive reactance is proportional to the operating frequency and the amount of inductance. Doubling the frequency, for example, doubles the amount of inductive reactance. Or if you want to decrease the inductive reactance at a certain frequency, you have to decrease the amount of inductance by a proportional amount.

By way of a numerical example, suppose that you are operating a 1-H inductor in a 100-Hz circuit. How much inductive reactance does the coil provide?

$$X_L = 2\pi f L$$
$$X_L = 2 \times 3.14 \times 100 \text{ Hz} \times 1 \text{ H}$$
$$X_L = 628 \text{ }\Omega$$

So a 1-H inductor in a 100-Hz circuit offers 628 Ω of inductive reactance. How much inductive reactance will you find when operating a 0.1-H coil in a dc circuit? (Hint: the frequency of a dc circuit is 0 Hz.)

APPLICATIONS OF INDUCTANCE

A filter circuit can use inductive reactance to pass some frequencies while blocking others.

Because the amount of inductive reactance depends on frequency, inductors are often used in filter circuits. "Filters" are special circuits that have the property of allowing certain frequencies to pass, while blocking others.

A low-pass filter circuit passes low frequencies, but blocks high frequencies.

Figure 11-9 shows an inductor and resistor connected as a low-pass filter. A "low-pass filter" passes low frequencies and blocks high frequencies. The inductor offers little inductive reactance to low frequencies, so low frequencies from the input voltage source pass easily to the resistor at the output of the circuit. Increasing the frequency at the input increases the inductive reactance of the inductor, thereby allowing a smaller portion of the waveform to reach the output resistor.

**Figure 11-9.
Simple Low-Pass Filter Circuit**

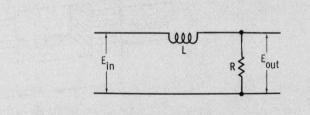

A high-pass filter circuit passes high frequencies, but blocks low frequencies.

Figure 11-10 shows the same two components—an inductor and resistor—connected as a high-pass filter. A "high-pass filter" passes high frequencies and blocks lower frequencies. The inductor responds the same way as in a low-pass filter, but it is now connected across the output terminals of the circuit. Located in that position, it bypasses lower frequencies and allows only higher frequencies to appear at the output terminals.

**Figure 11-10.
Simple High-Pass Filter
Circuit**

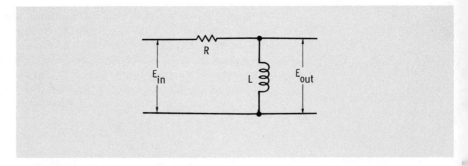

Inductors are used in high-quality audio amplifiers. High-pass and low-pass inductor circuits separate the low audio frequencies from the high frequencies. The low frequencies are then fed to larger loudspeakers (woofers) and the high frequencies are sent to smaller speakers (tweeters). Thus, each speaker reproduces only those frequencies that it can handle without significant distortion.

Transformers, too, rely heavily on inductive effects, including inductive reactance. Most transformers are constructed as shown in *Figure 11-11*. Two separate coils of wire are wound at different places on a common core, usually made of iron. An ac waveform applied to the primary winding produces magnetic lines of force in the core which create a voltage at the secondary winding. You will learn more about the operation of transformers in Chapter 18. What is important now is that you understand the role that inductive reactance plays in practical transformer devices.

**Figure 11-11.
Typical Transformer**

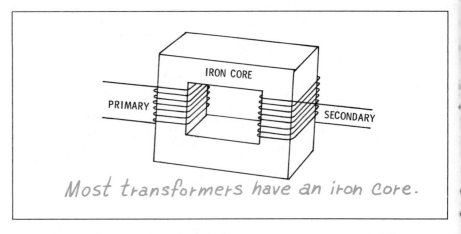

Most transformers have an iron core.

The concept of inductive reactance explains why you can leave the primary winding of a common doorbell transformer permanently connected to your household 120-V ac source without running up thousands of dollars in electric utility bills.

THE BASICS OF INDUCTANCE 11

If you measure the resistance of the transformer primary with an ohmmeter, you will most likely find a resistance of 1 Ω or less. Suppose, for the sake of argument, that you read exactly 1 Ω. How much current flows through a 1-Ω resistance connected to 120 V? By Ohm's law, I = E/R, the current will be 120 A. Indeed the 1-Ω transformer winding is connected to a 120-V source, but you certainly do not have 120 A flowing through it. That much current would blow a typical 15-A household fuse in no time. And even if you bypassed the fuse, that much current would blast that little transformer to bits.

Nothing seems to happen when you connect the doorbell to your 120-V ac lines. If you were to measure the ac current flowing through the primary coil, you would find it to be on the order of a few milliamperes. There seems to be a contradiction of facts here; but the matter can be resolved in the light of a few additional facts.

First, an ohmmeter uses a small amount of dc current to measure resistance. So measuring a coil (or transformer winding) with an ohmmeter shows the coil's opposition to that dc current flow. The dc resistance of a transformer winding can be quite low indeed. But its inductive reactance in an ac circuit can be quite high. Your 120-V ac utility service supplies a 60-Hz waveform. The inductive reactance of the transformer at 60 Hz is on the order of 10,000 ohms or more.

There are a couple of important lessons to be learned here. First, never connect a transformer to a dc source. Second, inductive reactance can play an important role in the proper operation of certain electrical devices, including transformers. And finally, what we see as an apparent contradiction of fact is often the result of our not knowing all the facts related to a problem.

Inductors are often used to modify the shape of pulse waveforms. When voltage pulses are applied to a circuit containing an inductor, the inductor opposes the change in current flow. As shown in *Figure 11-12*, the effect is to distort the waveform by rounding off the sharp corners on the leading and trailing edges.

**Figure 11-12.
Inductor and Pulsed
Current Waveform**

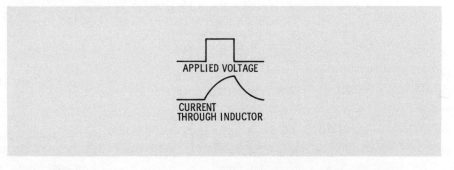

The voltage waveform across the inductor is quite different from the current waveform. *Figure 11-13* shows that an inductor tends to peak the edges of the voltage waveform.

11 THE BASICS OF INDUCTANCE

**Figure 11-13.
Inductor and Pulsed
Voltage Waveform**

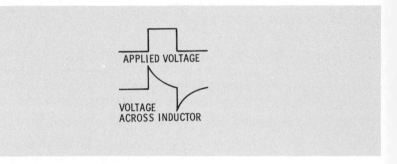

WHAT HAVE WE LEARNED?

1. Inductance is the result of an interaction between the magnetic field created by current flow and a counter emf created by the same magnetic field.
2. Inductance opposes a change in current flow.
3. The unit measurement for inductance is the henry (H).
4. Inductance consumes no electrical power; it stores energy as magnetic fields expand, then it returns the energy to the circuit when the fields are allowed to collapse.
5. Practical inductors are made of a coil of wire wrapped around a core material.
6. Inductors are sometimes called coils and chokes.
7. The value of inductance is determined by the number of turns of wire, the diameter of the coil of wire, and the material within the coil.
8. An inductance causes an ac sine wave of current to lag the applied voltage by 90°.
9. A coil's opposition to ac current flow is called inductive reactance.
10. The equation for inductive reactance is $X_L = 2\pi fL$.
11. The unit of measurement for inductive reactance is the ohm.
12. High-pass frequency filters can use inductors to pass higher frequencies while blocking lower frequencies.
13. Low-pass frequency filters can use inductors to pass lower frequencies while blocking higher frequencies.
14. Transformers rely on inductive reactance to limit the amount of ac current flow to a useful level.
15. An inductor rounds the edges of current pulses, but it peaks the edges of voltage pulses.

KEY WORDS

Choke
Coil
Counter emf
Henry (H)
High-pass filter

Inductance
Inductive reactance
Lagging waveform
Leading waveform
Low-pass filter

THE BASICS OF INDUCTANCE

Quiz for Chapter 11

1. Which one of the following statements best describes the nature of counter emf in an inductor?
   a. Counter emf tends to oppose any change in current flow through an inductor.
   b. Counter emf always reduces the amount of current flowing through an inductor.
   c. Counter emf always aids the current flowing through an inductor.
   d. Counter emf has no effect on the flow of ac current through an inductor.

2. The energy absorbed by an inductor in an ac circuit is stored as:
   a. heat.
   b. counter emf.
   c. magnetic lines of force.
   d. static electricity.

3. One henry of inductance is defined as:
   a. the amount of counter emf required to reduce a current to 1 A.
   b. the amount of inductance required for generating 1 V of counter emf when the current changes at the rate of 1 A per second.
   c. the number of turns of wire in an inductor multiplied by the amount of current flowing through it.
   d. the amount of inductance required to change the frequency of a current by 1 Hz.

4. Doubling the number of turns of wire in an inductor:
   a. multiplies the value of inductance by two.
   b. reduces the value of inductance by one-half.
   c. multiplies the value of inductance by four.
   d. reduces the value of inductance by one-fourth.
   e. has no effect on the inductance.

5. Doubling the diameter of the coil in an inductor:
   a. multiplies the value of inductance by two.
   b. reduces the value of inductance by one-half.
   c. multiplies the value of inductance by four.
   d. reduces the value of inductance by one-fourth.
   e. has no effect on the inductance.

6. Reducing the amount of current through an inductor:
   a. multiplies the value of inductance by two.
   b. reduces the value of inductance by one-half.
   c. multiplies the value of inductance by four.
   d. reduces the value of inductance by one-fourth.
   e. has no effect on the inductance.

11 THE BASICS OF INDUCTANCE

7. Which one of the following statements most accurately describes the nature of inductive reactance?
   a. Inductive reactance is an opposition to current flow in dc circuits.
   b. Inductive reactance is an opposition to current flow in ac circuits.
   c. Inductive reactance is a force that tends to overcome the effects of counter emf.
   d. Inductive reactance is another term for counter emf.

8. Inductive reactance is measured in units of:
   a. volts per second.
   b. amperes per second.
   c. ohms.
   d. henrys.

9. Which one of the following statements regarding ac circuits is correct?
   a. Current and voltage are in phase through a resistor.
   b. Current leads the voltage through a resistor.
   c. Current lags the voltage through a resistor.

10. Which one of the following statements regarding ac circuits is correct?
    a. Current and voltage are in phase through an inductor.
    b. Current leads the voltage through an inductor.
    c. Current lags the voltage through an inductor.

11. What is the inductive reactance of a 2-H coil operating at 60 Hz?
    a. 0.033 A.
    b. 30 Ω.
    c. 60 Ω.
    d. 120 Ω.
    e. 754 Ω.

12. A low-pass filter:
    a. passes lower-frequency signals and reduces higher-frequency signals.
    b. passes higher-frequency signals and reduces lower-frequency signals.
    c. reduces the level of all frequencies by a significant amount.
    d. has no effect on ac signals.

13. A high-pass filter:
    a. passes lower-frequency signals and reduces higher-frequency signals.
    b. passes higher-frequency signals and reduces lower-frequency signals.
    c. increases the level of all frequencies by a significant amount.
    d. has no effect on ac signals.

Inductive Circuits

ABOUT THIS CHAPTER

In the previous chapter we examined the basic elements of inductance. You learned that an inductor attempts to oppose any change in current flow through it, called inductive reactance. In this chapter, you will learn more about inductors and inductive circuits, including how to calculate the total inductance and total inductive reactance of various combinations of inductors—series, parallel, and combinations of series and parallel circuits. Beyond that, we will calculate the current through inductors and the voltage drop across them.

A PURELY INDUCTIVE CIRCUIT

In a purely inductive circuit, the only opposition to current flow is the inductive reactance of the inductors.

A purely inductive circuit contains only inductances and the only opposition to current flow is provided by inductive reactance. As shown in the vector diagram in *Figure 12-1*, the current through a purely inductive circuit lags the applied ac voltage by exactly 90°.

Figure 12-1.
Current Lags Voltage in Purely Inductive Circuit

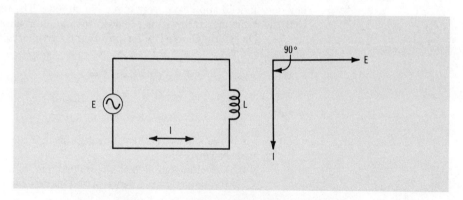

We know that the inductive reactance of an inductor is determined by the value of inductance and the operating frequency of the ac source. The mathematical expression for this relationship is:

$$X_L = 2\pi f L$$

12 INDUCTIVE CIRCUITS

where,

X_L is the inductive reactance in ohms,
f is the operating frequency in hertz,
L is the value of the inductor in henrys,
π is the constant 3.14.

In reality, there is no such thing as a purely inductive circuit. The concept of a pure inductor, however, is an important one that you will use for analyzing the operation of practical circuits that combine inductance, resistance, and (as you will learn later) capacitance.

Like resistors, two or more inductors can be connected with one another in a variety of ways. Inductors can be connected in series, parallel, and combinations of series and parallel.

OHM'S LAW APPLIED TO INDUCTIVE CIRCUITS

The basic form of Ohm's law states that the amount of current flowing through a circuit is equal to the applied voltage divided by the opposition to current flow. According to this important principle, larger values of voltage cause larger amounts of current to flow. On the other hand, increasing the amount of opposition to current flow decreases the amount of current.

When you are working with circuits that contain nothing but pure resistances, the resistors provide the opposition to current flowing through the circuits. When working with purely inductive circuits, the inductive reactance of the inductors provides the opposition to current flow. Ohm's law does indeed apply to purely inductive circuits. The only difference is that the resistance term (R) is replaced with inductive reactance (X_L).

Therefore, Ohm's law for purely inductive circuits looks like this:

$$I = \frac{E}{X_L}$$

where,

I is the amount of current flowing through the inductance in amperes (usually rms)
E is the voltage applied to the inductor in volts (usually rms),
X_L is the inductive reactance of the inductor in ohms.

Figure 12-2 shows a circuit composed of an ac power source and a single inductor. The source is rated at 12 V ac, 100 Hz, and the value of the inductor is 0.1 H. What is the current through this circuit? The inductive-reactance version of Ohm's law provides a means for calculating the current, but first we must determine the value of inductive reactance.

$$X_L = 2\pi fL$$
$$X_L = 2 \times 3.14 \times 100 \text{ Hz} \times 0.1 \text{ H}$$

Ohm's law applied to purely inductive circuits: $I = E/X_L$

INDUCTIVE CIRCUITS

12

$$X_L = 6.28 \times 10$$
$$X_L = 62.8 \, \Omega$$

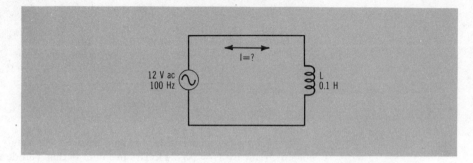

Figure 12-2. Determining Current through Simple Inductive Circuit

Now that you know that the amount of inductive reactance is 62.8 Ω, you can apply the inductive-reactance version of Ohm's law to determine the current through the inductor.

$$I = \frac{E}{X_L}$$
$$I = \frac{12 \text{ V}}{62.8}$$
$$I = 0.19 \text{ A}$$

Table 12-1 summarizes three forms of the basic inductance equation and Ohm's law as it applies to ideal inductive circuits. To use the equations in the table, consider the information supplied for the inductor circuit in *Figure 12-3*. You know the applied ac voltage, operating frequency, and the desired amount of ac current. The problem is to determine the value of inductance required.

Table 12-1. Basic Inductance Equation and Ohm's Law

Inductance Equations	Ohm's Law Equations (pure inductance)
$X_L = 2\pi f L$	$I = E/X_L$
$L = X_L/2\pi f$	$E = IX_L$
$f = X_L/2\pi L$	$X_L = E/I$

X_L = inductive reactance in ohms
L = inductance in henrys
f = frequency in hertz
I = current in amperes (usually rms)
E = voltage in volts (usually rms)

UNDERSTANDING ELECTRICITY AND ELECTRONICS CIRCUITS

12 INDUCTIVE CIRCUITS

**Figure 12-3.
Determining Inductance**

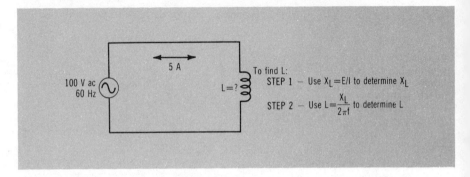

Table 12-1 shows that you can calculate the value of inductance in such a circuit by $L = X_L/2\pi f$. The problem is that you do not yet know the value of inductive reactance, X_L. You begin by using the form of Ohm's law that provides the amount of inductive reactance in terms of voltage and current:

$$X_L = \frac{E}{I}$$

$$X_L = \frac{100}{5}$$

$$X_L = 20 \; \Omega$$

Knowing the value of inductive reactance, you can now apply the equation that solves for the value of inductance:

$$L = \frac{X_L}{2\pi f}$$

$$L = \frac{20}{(6.28 \times 60)}$$

$$L = 0.053 \; H$$

An inductor having a value of 0.052 H solves the problem.

DC POWER EQUATIONS APPLIED TO INDUCTIVE CIRCUITS

The versions of Ohm's law that you studied in connection with resistances in ac and dc circuits apply equally well to ideal inductive circuits. You have just seen that you can simply replace resistance (R) with inductive reactance (X_L) in the basic Ohm's law equations. However, this is not true for the power equations you learned in connection with resistors in ac and dc circuits. An explanation is in order.

Figure 12-4 shows the sine waveforms for current, voltage, and power in a resistor circuit—where the only opposition to current flow is from resistances. Notice that the voltage and current waveforms are in phase, an important characteristic of resistor circuits. If you apply the power equation, $P = IE$, at every point along the voltage and current

INDUCTIVE CIRCUITS

12

waveforms, you find that the power waveform looks like a series of humps that are always positive (multiplying two positive values or two negative values always produces a positive result).

**Figure 12-4.
Voltage, Current, and Power Waveforms for Purely Resistive Circuits**

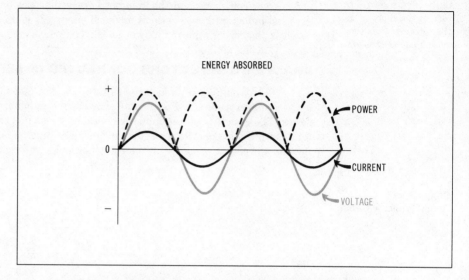

You have already learned that no power is dissipated in an ac circuit that contains only inductance. Observing the waveforms in *Figure 12-5* will help you understand this fact. First, notice that the current waveform lags the voltage waveform. In an ideal inductive circuit, the current lags the voltage by exactly 90°. If you apply the power equation (P = IE) every step along the way, you find that the power waveform is also a sine waveform that is positive half the time and negative half the time.

**Figure 12-5.
Voltage, Current, and Power Waveforms for Purely Inductive Circuits**

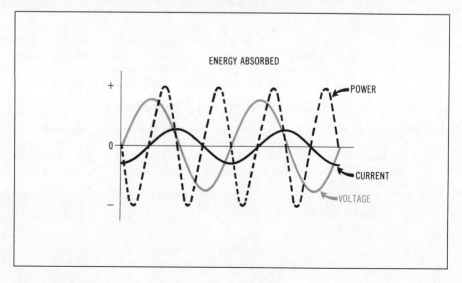

12 INDUCTIVE CIRCUITS

The power equations have no practical meaning for purely inductive circuits. Such circuits always return as much power as they absorb. Thus the overall power consumption is zero.

During the time the power waveform is positive, the inductor absorbs power from the circuit in order to build up its magnetic lines of force. During the time the power waveform is negative, the magnetic lines of force are collapsing and the inductor is returning power to the circuit. An ideal inductor always returns as much energy as it uses so the overall result is that it consumes no power at all. The power equation has no real meaning when applied to an ideal inductive circuit.

ANALYZING INDUCTORS CONNECTED IN SERIES

Any number of inductors can be connected end to end to make up a series inductor circuit. *Figure 12-6* shows three inductors connected in series. The circuit is purely inductive because it has no other component that can provide opposition to current flow. The only opposition to current flow in this circuit comes from the combined inductive reactance of the inductors.

**Figure 12-6.
Series Inductor Circuit**

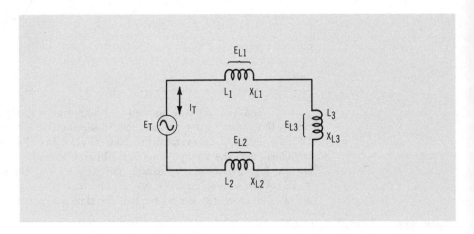

Like any other kind of series circuit, the same current flows through all parts of a series inductor circuit. And if the circuit is purely inductive (as indicated in this example), the source voltage is divided among the inductors according to their values of reactance.

Determining Total Inductance

The total inductance of two or more inductors connected in series is equal to the sum of the individual inductor values.

The total inductance of two or more inductors connected in series is equal to the sum of their individual inductances. If the total inductance is shown as L_T and the individual inductance values as L_1, L_2, L_3, and so on, the equation for determining the total inductance of a series-inductor circuit is:

$$L_T = L_1 + L_2 + L_3\ldots$$

A series circuit composed of three inductors rated at 1 H, 4 H, and 3 H has a total inductance of $1 + 4 + 3$, or 8 H.

12 INDUCTIVE CIRCUITS

Determining Inductive Reactances in a Series Circuit

Any inductor offers some opposition to ac current flow through a circuit. In a purely inductive circuit, the opposition to current flow takes the form of inductive reactance.

In a series inductor circuit, the total inductive reactance is equal to the sum of the reactances of the individual inductors.

Each inductor in a series inductor circuit contributes its share of reactance, or opposition, to current flow. The total inductive reactance of a series circuit is equal to the sum of the individual reactances. Using X_{LT} as the mathematical symbol for total inductive reactance and X_{L1}, X_{L2}, X_{L3}, and so on, as the reactances for inductors L_1, L_2, and L_3, a mathematical expression for total inductive reactance is:

$$X_{LT} = X_{L1} + X_{L2} + X_{L3}...$$

A series circuit composed of three inductors having reactances of 100 Ω, 40 Ω, and 55 Ω has a total inductive reactance of 100 + 40 + 55, or 195 Ω.

Now suppose that you are working with the circuit shown in *Figure 12-7* and you want to know its total inductive reactance. You know that the total inductive reactance will be the sum of the reactances of the two inductors. That information is not provided directly, though. So you have to calculate the individual inductive reactances first. In this instance, X_{L1} is close to 251 Ω at 400 Hz, and X_{L2} is close to 1005 Ω at 400 Hz. The sum of the two values provides the total inductive reactance of the circuit, approximately 1256 Ω.

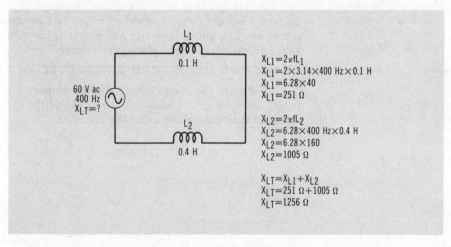

Figure 12-7. Total Inductive Reactance of Series Inductor Circuit

Determining Voltage Drops in Series Inductive Circuits

You can always determine the voltage drop across an inductor in an ac circuit by applying an inductive-reactance form of Ohm's law, $E = IX_L$. That formula can be applied to any inductor in a series circuit, assuming that you know the amount of current flowing through the circuit and the inductive reactance of the inductor in question.

So if inductor L_1 in a series circuit has an inductive reactance of 250 Ω and the current flowing through the circuit is 0.1 A, the voltage drop across that inductor is:

$$E_{L1} = IX_{L1}$$
$$E_{L1} = 0.1 \times 250$$
$$E_{L1} = 25 \text{ V}$$

In most practical situations, however, you are provided with the circuit's source voltage, the operating frequency, and the inductance values of the individual inductors. None of that information fits directly into the inductive-reactance version of Ohm's law for voltage. This means you must use the information at hand to determine the inductive reactance of each inductor and the amount of current flowing through the circuit. *Figure 12-8* illustrates this procedure for a specific circuit.

With this information, your first step is to determine the inductive reactance of each inductor in the circuit. The second step is to add the reactances to find the total inductive reactance. Knowing the total inductive reactance lets you apply Ohm's law to determine the current through the circuit—the same current that flows through all of these series-connected inductors. Finally, knowing the current and inductive reactance for each inductor, you can apply Ohm's law to calculate the voltage drop across each inductor.

If you sum the voltage drops across each inductor in this example, you will find that they very nearly total the source voltage. The small difference is mainly due to round-off errors that accumulated during the procedure. In a purely inductive series circuit, the sum of the individual voltage drops should equal the source voltage.

In a purely inductive series circuit, the sum of the voltage drops across each inductor is equal to the source voltage.

ANALYZING INDUCTORS CONNECTED IN PARALLEL

Figure 12-9 shows three inductors connected in parallel. It is a purely inductive circuit because the only opposition to current flow is the inductive reactance of the individual branches.

INDUCTIVE CIRCUITS

Figure 12-8.
Voltage Drops across Inductive Reactances in Series

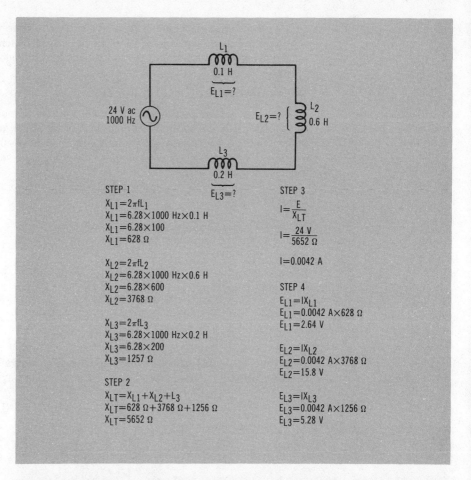

Figure 12-9.
Parallel Inductor Circuit

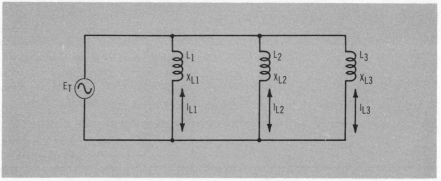

12 INDUCTIVE CIRCUITS

Like any other kind of parallel circuit, the same voltage appears across all parts of a parallel inductor circuit. And if the circuit is purely inductive (as in this example), the source current is divided among the inductors according to their values of reactance.

Determining the Total Inductance

The total inductance of two or more inductors connected in parallel is less than the value of the smallest inductor.

You determine the total inductance of two or more inductors connected in parallel by applying equations of the same form you use for finding the total resistance of a parallel circuit. As a result, the total inductance of a parallel inductor circuit is less than the value of the smallest inductor.

The general equation for calculating the total inductance of a series of inductors connected in parallel is:

$$L_T = \frac{1}{[(1/L_1) + (1/L_2) + (1/L_3)\ldots]}$$

But if you prefer to work with two inductors at a time, you can apply the simpler product-over-sum rule:

$$L_T = \frac{(L_1 \times L_2)}{(L_1 + L_2)}$$

For example, you calculate the total inductance of a parallel circuit composed of two inductors rated at 1 H and 4 H this way:

$$L_T = \frac{(L_1 \times L_2)}{(L_1 + L_2)}$$
$$L_T = \frac{(1 \times 4)}{(1 + 4)}$$
$$L_T = \frac{4}{5}$$
$$L_T = 0.8 \text{ H}$$

Figure 12-10 shows how you can use the product-over-sum rule to determine the total inductance of a circuit having three inductors connected in parallel. The first step deals with inductors L_2 and L_3 to determine the value of an equivalent inductance, L_A. The second step combines L_1 and L_A to provide the total inductance of the circuit.

Determining Inductive Reactances in a Parallel Circuit

Ac current flows through each branch of a parallel inductive circuit. The amount of current in each branch depends on the inductive reactance of that branch—the larger the amount of inductive reactance, the smaller the amount of current.

INDUCTIVE CIRCUITS

**Figure 12-10.
Total Inductance by
Product-Over-Sum Rule**

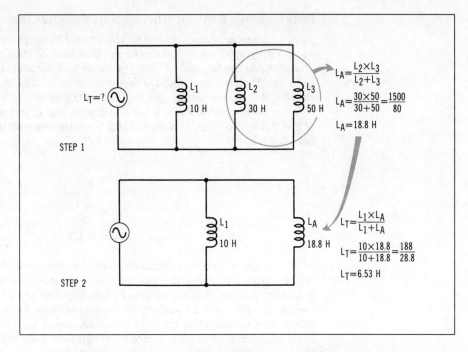

Use the product-over-sum rule to determine the total inductive reactance of two inductors connected in parallel.

The equations for determining the total reactance of a parallel circuit have the same general form as those for determining the total inductance. The general equation is:

$$X_{LT} = \frac{1}{[(1/X_{L1}) + (1/X_{L2}) + (1/X_{L3})\ldots]}$$

and the product-over-sum version is:

$$X_{LT} = \frac{(X_{L1} \times X_{L2})}{(X_{L1} + X_{L2})}$$

Use the product-over-sum rule to confirm that a parallel circuit composed of two inductors having reactances of 100 Ω and 800 Ω has a total inductive reactance of 44.4 Ω.

Many practical situations require that you determine the total inductive reactance of a parallel circuit, without knowing the amounts of reactance. In such situations, you may be given a source voltage, the operating frequency, and the inductance of the individual coils. The general procedure is to use the given values of frequency and inductances to calculate the inductive reactance of each inductor, and then to determine the total reactance from your results.

Determining Branch Currents in Parallel Inductive Circuits

You can always determine the current through an inductor in an ac circuit by applying an inductive-reactance form of Ohm's law: $I = E/X_L$. That formula can be applied to any inductor in a parallel circuit, assuming you know the amount of voltage dropped across it. And in simple parallel circuits, the voltage across each inductor is the same as the source voltage.

So if inductor L_1 in a parallel circuit has an inductive reactance of 250 Ω and the source voltage is 100 V ac, the current through the inductor is:

$$I_{L1} = \frac{E}{X_{L1}}$$

$$I_{L1} = \frac{100}{250}$$

$$I_{L1} = 0.4 \text{ A}$$

In most practical situations, however, you have the source voltage, the operating frequency, and the inductance values of the individual inductors. None of this information fits directly into the inductive-reactance version of Ohm's law for current. So you should use the available information to determine the inductive reactance of each inductor, and then apply Ohm's law to find the current through each inductor. *Figure 12-11* illustrates the procedure.

In a purely inductive parallel circuit, the total circuit current is equal to the sum of the currents through each branch.

In a purely inductive parallel circuit, the sum of the currents through each inductor is equal to the source current. The total current can be determined by Ohm's law—dividing the source voltage by the total inductive reactance.

ANALYZING COMBINATION INDUCTIVE CIRCUITS

You have already learned how to analyze combination circuits made up of resistors. The procedures for analyzing purely inductive combination circuits are practically identical. The only difference is that you must be prepared to calculate values of inductive reactance for the inductive versions of such circuits.

INDUCTIVE CIRCUITS

**Figure 12-11.
Currents through
Inductive Reactances in
Parallel**

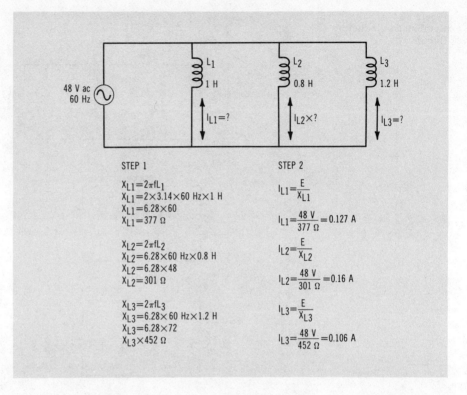

STEP 1

$X_{L1} = 2\pi f L_1$
$X_{L1} = 2 \times 3.14 \times 60 \text{ Hz} \times 1 \text{ H}$
$X_{L1} = 6.28 \times 60$
$X_{L1} = 377 \ \Omega$

$X_{L2} = 2\pi f L_2$
$X_{L2} = 6.28 \times 60 \text{ Hz} \times 0.8 \text{ H}$
$X_{L2} = 6.28 \times 48$
$X_{L2} = 301 \ \Omega$

$X_{L3} = 2\pi f L_3$
$X_{L3} = 6.28 \times 60 \text{ Hz} \times 1.2 \text{ H}$
$X_{L3} = 6.28 \times 72$
$X_{L3} \times 452 \ \Omega$

STEP 2

$I_{L1} = \dfrac{E}{X_{L1}}$

$I_{L1} = \dfrac{48 \text{ V}}{377 \ \Omega} = 0.127 \text{ A}$

$I_{L2} = \dfrac{E}{X_{L2}}$

$I_{L2} = \dfrac{48 \text{ V}}{301 \ \Omega} = 0.16 \text{ A}$

$I_{L3} = \dfrac{E}{X_{L3}}$

$I_{L3} = \dfrac{48 \text{ V}}{452 \ \Omega} = 0.106 \text{ A}$

Figure 12-12 shows a simple combination circuit composed of three inductors. The circuit shows the source voltage and frequency and the values of the individual inductors. When you complete an analysis of this circuit, you will know the total circuit current and reactance and the reactance, current, and voltage drop for each inductor.

12 INDUCTIVE CIRCUITS

Figure 12-12.
Combination Inductor Circuit

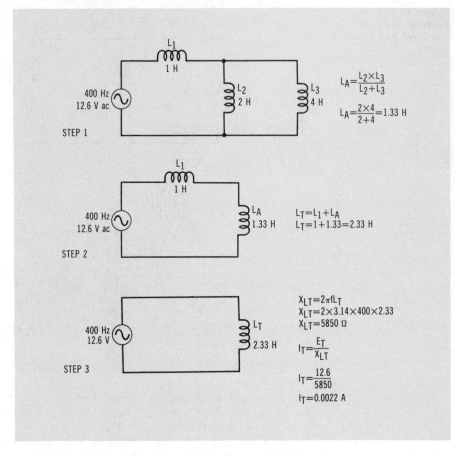

The general approach to analyzing such circuits is the same as the approach for analyzing resistor circuits: do what you can, when you can do it. In this example, the first thing you can do is combine the inductances of parallel inductors L_2 and L_3 to arrive at an equivalent inductance value, L_A. This is shown in Step 1 of *Figure 12-12*.

The next thing you can do is to combine the series arrangement of inductors L_1 and L_A as shown in Step 2, giving the total inductance of the circuit, 1.33 H.

In Step 3, you use the information at hand to determine the total inductive reactance of the circuit and the total current. Inductor L_1 is connected in series with the source, so you know that the current through that inductor is the same as the total circuit current—0.002 A. That same amount of current flows through the equivalent inductance, L_A, as shown in Step 4. Why is this so? Because inductors L_1 and L_A make up a simple series circuit, and currents are the same through series-connected components.

INDUCTIVE CIRCUITS 12

Figure 12-12. cont.

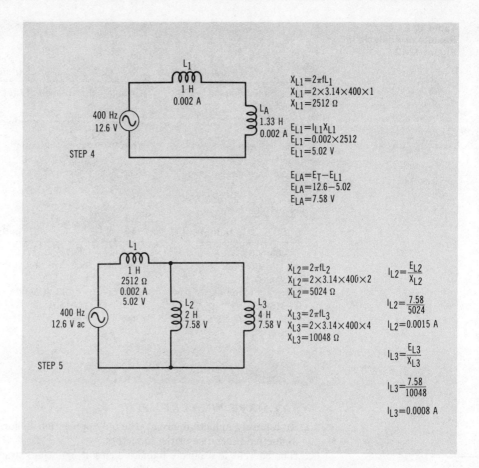

In Step 4, you first determine the inductive reactance of L_1, and then use Ohm's law to find the voltage drop across that inductor—multiplying its current by its reactance. Knowing that in purely inductive series circuits voltages add up to the source voltage, you can subtract the voltage across L_1 from the source voltage to find the voltage across the equivalent inductance, L_A.

Step 5 uses the information you have gathered to complete the analysis for L_1 and uses the inductive-reactance formula and Ohm's law to complete the analysis for the parallel arrangement of L_2 and L_3.

A complete summary of this analysis is shown in *Table 12-2*.

UNDERSTANDING ELECTRICITY AND ELECTRONICS CIRCUITS

**Table 12-2.
Results of Analysis in
Figure 12-12**

Total Circuit		Inductor L_1	
Total circuit voltage: $E_T =$ 12.6 V	(Given)	Inductance of L_1: $L_1 = 1$ H	(Given)
Operating frequency: $f =$ 400 Hz	(Given)	Inductive reactance of L_1: $X_{L1} = 2512$ Ω	(Step 3)
Total inductance: $L_T =$ 1.33 H	(Step 2)	Current through L_1: $I_{L1} =$ 0.002 A	(Step 4)
Total inductive reactance: $X_{LT} = 5850$ Ω	(Step 3)	Voltage across L_1: $E_{L1} =$ 5.02 V	(Step 4)
Total circuit current ($I_T =$ 0.0022 A	(Step 3)		
Inductor L_2		**Inductor L_3**	
Inductance of L_2: $L_2 = 2$ H	(Given)	Inductance of L_3: $L_3 = 4$ H	(Given)
Inductive reactance of L_2: $X_{L2} = 5024$ Ω	(Step 5)	Inductive reactance of L_3: $X_{L3} = 10048$ Ω	(Step 5)
Current through L_2: $I_{L2} =$ 0.0015 A	(Step 5)	Current through L_3: $I_{L3} =$ 0.0015 A	(Step 5)
Voltage across L_2: $E_{L2} =$ 7.58 V	(Step 4)	Voltage across L_3: $E_{L1} =$ 7.58 V	(Step 4)

WHAT HAVE WE LEARNED?

1. In a purely inductive circuit, the only opposition to current flow is the inductive reactance of the inductors.
2. The current in a purely inductive circuit lags the voltage by exactly 90°.
3. Although there is no such thing as a pure inductor or a purely inductive circuit, the principles are important to the analysis of ac circuits that include inductors.
4. Ohm's law applies to purely inductive circuits; the opposition to current flow is inductive reactance rather than resistance.
5. Dc power equations have no meaning in purely inductive circuits.
6. The total inductance of inductors connected in series is equal to the sum of the individual inductance values.
7. The same current flows through all inductors in a series circuit.
8. In a purely inductive series circuit, the sum of the voltage drops across the inductors is equal to the source voltage.
9. The sum of the individual inductive reactances in a series circuit is equal to the total inductive reactance of the circuit.
10. Inductors connected in parallel have the same voltage dropped across them.

INDUCTIVE CIRCUITS

11. You can use the product-over-sum rule to determine the total inductance of two inductors connected in parallel. The total inductance of such a circuit is less than the value of the smaller inductance.
12. In a purely inductive circuit, the total current through a parallel section of the circuit is equal to the sum of the currents in the individual branches.
13. You can use the product-over-sum rule to determine the total inductive reactance of two inductors connected in parallel. The total inductive reactance of such a circuit is always less than the value of the smaller reactance.

12 INDUCTIVE CIRCUITS

Quiz for Chapter 12

1. Which one of the following phrases most accurately describes a purely inductive circuit?
   a. Resistances provide the only opposition to current flow.
   b. Inductive reactance provides the only opposition to current flow.
   c. Combinations of resistance and inductive reactance provide any opposition to current flow.
   d. The ac voltage lags the current by 90°.
   e. The ac current and voltage are in phase.

2. In a pure inductor:
   a. ac current and voltage are exactly in phase.
   b. ac current leads the voltage by 90°.
   c. ac current lags the voltage by 90°.
   d. ac current is converted to dc voltage.

3. Which one of the following statements is true?
   a. Inductive reactance is at a maximum for dc source voltages.
   b. Inductive reactance decreases with operating frequency.
   c. Inductive reactance increases with operating frequency.
   d. There is no meaningful relationship between inductive reactance and operating frequency.

4. What is the inductive reactance of a 0.1-H coil that is operating at 1000 Hz?
   a. Less than 1 Ω.
   b. 100 Ω.
   c. 314 Ω.
   d. 628 Ω.
   e. 1000 Ω.

5. How much current flows through a 0.02-H choke that is operating from a 12-V ac, 100-Hz source?
   a. 0.02 A.
   b. 0.995 A.
   c. 2.02 A.
   d. 10 A.
   e. 100 A.

6. Doubling the operating frequency of an inductive circuit:
   a. has no effect on the inductive reactance.
   b. doubles the amount of inductive reactance.
   c. cuts the inductive reactance in half.
   d. multiplies the inductive reactance by 6.28.

7. Doubling the operating frequency of a purely inductive circuit:
   a. has no effect on the current through the inductors.
   b. doubles the amount of current through the inductors.
   c. cuts the current through the inductors by one-half.
   d. increases the current, but by an amount that can be determined only by doing a complete analysis of the circuit.
   e. decreases the current, but by an amount that can be determined only by doing a complete analysis of the circuit.

INDUCTIVE CIRCUITS

12

8. Doubling the operating voltage of a purely inductive circuit:
   a. has no effect on the inductive reactance.
   b. doubles the amount of inductive reactance.
   c. cuts the inductive reactance in half.
   d. multiplies the inductive reactance by 6.28.

9. Which one of the following statements is true?
   a. Power dissipation of a pure inductor is at a maximum for dc source voltages.
   b. Power dissipation of a pure inductor decreases with operating frequency.
   c. Power dissipation of a pure inductor increases with operating frequency.
   d. There is no meaningful relationship between the power dissipation of a pure inductor and its operating frequency.

10. The total inductance of a series inductor circuit is:
    a. equal to the sum of the individual inductance values.
    b. equal to the sum of the individual inductive-reactance values.
    c. equal to the source voltage divided by total current.
    d. less than the value of the smallest inductor.

11. The total inductive reactance of a parallel inductor circuit is:
    a. equal to the sum of the individual inductance values.
    b. equal to the sum of the individual inductive-reactance values.
    c. equal to the source voltage divided by total current.
    d. less than the inductance value of the smallest inductor.

12. What is the voltage drop across inductor L_1 in *Figure 12-13*?
    a. 40 V.
    b. 100 V.
    c. 200 V.
    d. 300 V.
    e. 400 V.

Figure 12-13.
Circuit for Questions 12-16

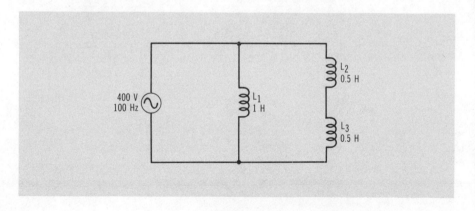

UNDERSTANDING ELECTRICITY AND ELECTRONICS CIRCUITS

INDUCTIVE CIRCUITS

13. What is the voltage drop across inductor L_2 in *Figure 12-13*?
  a. 20 V.
  b. 50 V.
  c. 150 V.
  d. 200 V.
  e. 300 V.
  f. 400 V.

14. What is the voltage drop across inductor L_3 in *Figure 12-13*?
  a. 20 V.
  b. 50 V.
  c. 150 V.
  d. 200 V.
  e. 300 V.
  f. 400 V.

15. What is the current through inductor L_1 in *Figure 12-13*?
  a. 314 mA.
  b. 628 mA.
  c. 637 mA.
  d. 1.27 A.
  e. 4 A.

16. What is the current through inductor L_3 in *Figure 12-13*?
  a. 157 mA.
  b. 314 mA.
  c. 628 mA.
  d. 637 mA.
  e. 1.27 A.
  f. 2 A.

13

Understanding RL Circuits

ABOUT THIS CHAPTER

You have just learned how you can analyze purely inductive circuits. In this chapter, we deal with circuits that are not purely inductive. You will learn more about phase angles, and you will calculate impedance and see how a resistance inserted into an inductive circuit affects the distribution of currents and voltages.

RL CIRCUITS

An RL circuit contains both resistance and inductance.

In Chapters 11 and 12, we described inductance and inductive circuits in terms of pure inductances. In such a circuit, inductive reactance provides the only opposition to current flow and ac current lags the applied voltage by exactly 90°. When you include a resistance in an inductor circuit, the circuit is no longer purely inductive—resistance and inductive reactance both provide opposition to current flow, and the ac current through the circuit lags the applied voltage by some amount less than 90°. A circuit that includes both resistance (R) and inductance (L) is called an RL circuit.

Figure 13-1 illustrates examples of RL circuits.

We use what we know about pure inductors to analyze RL circuits, but RL circuits require more extensive analysis.

SERIES RL CIRCUITS

Series RL circuits share many important characteristics with other kinds of series circuits. One of these is that the same current flows through every part of the circuit. But there are significant differences as well. For instance, the sums of the voltage drops across the components in a series RL circuit do not equal the source voltage.

The amplitude and phase of currents are identical in all parts of a series RL circuit.

The same current flows through every component in a series circuit. This means the amount of current (amplitude) through every component is equal to the amplitude of the total current:

$$I_T = I_L = I_R$$

where,

I_T is the total current in amperes,
I_L is the current through the inductance in amperes,
I_R is the current through the resistances in amperes.

13 UNDERSTANDING RL CIRCUITS

**Figure 13-1.
RL Circuits**

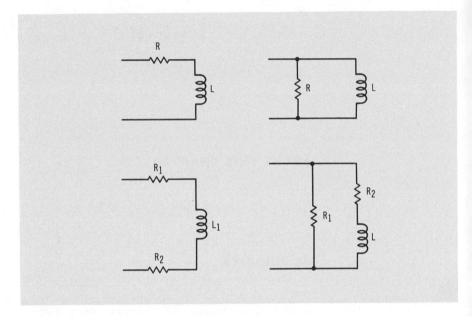

    The fact that the same current flows through every component in a series circuit means that the phase of the current is also the same through each component. *Figure 13-2* illustrates this. The sine-wave diagrams show that the currents through the resistance and inductance have the same amplitude and phase as the total current (the source current in this instance). The current vector diagram illustrates the same principle: the current vectors all have the same length and direction—they are superimposed on one another.

13

UNDERSTANDING RL CIRCUITS

**Figure 13-2.
Currents in Series RL Circuits**

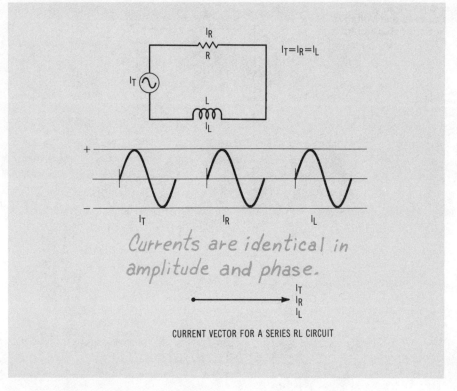

Voltage in a Series RL Circuit

The amplitude of the voltage drop across a resistance is equal to the amplitude of the current multiplied by the value of the resistance. Furthermore, the voltage and current waveforms are in phase, as shown in *Figure 13-3*.

**Figure 13-3.
Currents and Voltages
in Series RL Circuits**

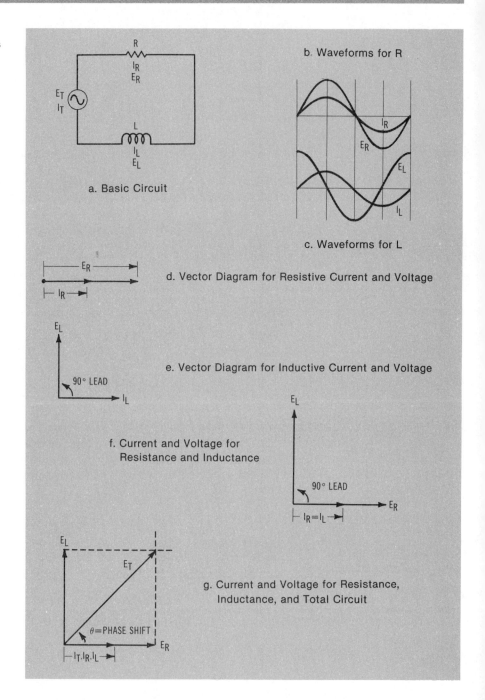

13 UNDERSTANDING RL CIRCUITS

You are already familiar with the following mathematical expression for the voltage across a resistance.

$$E_R = I_R \times R$$

where,

E_R is the amplitude of the voltage across the resistance in volts,
I_R is the amplitude of the current through the resistance in amperes,
R is the value of the resistance in ohms.

The voltage drop across an inductance is equal to the amplitude of the current through the inductance multiplied by the value of inductive reactance. The current through the inductance, however, lags its voltage drop by exactly 90°.

The equation for the voltage drop across an inductor is:

$$E_L = I_L \times X_L$$

where,

E_L is the amplitude of the voltage across the inductance in volts,
I_L is the amplitude of the current through the inductance in amperes,
X_L is the inductive reactance of the inductance in ohms.

It is often more helpful to express this principle in a somewhat different way: the voltage across an inductance leads the current by 90°. Like saying a glass is half-empty or half-full, the meaning is the same, only the perspective is different. When working with RL circuits, you will find it more convenient to say that the voltage across an inductance leads the current by 90°. The waveforms for the inductance in *Figure 13-3* reflect that perspective.

The voltage across the inductor in a series RL circuit leads the voltage across the resistance by 90°.

Now compare the waveforms in *Figure 13-3* for the resistance and inductance for a series RL circuit. You can see that the currents are in phase. That is a fixed principle for any kind of series circuit. The voltages, however, are 90° out of phase. The voltage across the inductance leads the voltage across the resistance by 90°.

The separate vector diagrams for the resistance and inductance show the phase relationships for those components. The vector diagram for the resistance shows the voltage and current vectors pointing in the same direction, indicating that they are in phase. The vector diagram for the inductance shows the inductive voltage leading the inductive current by 90°.

Because the amplitude and phase of the current through the resistance and inductance are the same in a series circuit, it is possible to combine the separate vector diagrams into one by aligning the current vectors for the two diagrams. The result, *Figure 13-3f*, is a vector diagram for the current and voltage in a series RL circuit. It shows the currents

13 UNDERSTANDING RL CIRCUITS

The total voltage in a series RL circuit is not equal to the sum of the resistive voltage and inductive voltage.

having the same amplitude and phase, the voltage across the resistor having the same phase as the currents, and the voltage across the inductance leading by 90°.

The fact that the inductive voltage leads the resistive voltage by 90° suggests another principle: the total voltage in a series RL circuit is *not* the simple sum of the resistive and inductive voltage. How do we represent the total voltage on a series RL vector diagram? If you regard the voltage vectors for the resistance and inductance as adjacent sides of a rectangle, you can complete the figure of a rectangle as shown in *Figure 13-3g*. Having completed the rectangle, the vector for the total voltage is represented by the diagonal of the rectangle. The length of the diagonal represents the amplitude of the total voltage, and the angle with respect to the current vectors represents the amount of phase shift between total current and total voltage.

The series RL vector diagram in *Figure 13-4* holds the clue to determining the amplitude of the total voltage and also the phase angle between total voltage and total current. If you are not familiar with calculating square roots and trigonometry, you can determine the length of the E_T vector and the phase angle by graphical methods. Begin by carefully plotting the vectors for E_T and E_L on a sheet of graph paper. Plot these two vectors at right angles as shown in the figure, and give them lengths that are proportional to their amplitudes. If $E_R = 4$ V and $E_L = 3$ V, for example, plot the E_R vector four units long and the E_L vector three units long. When you've plotted the E_R and E_L vectors, complete the rectangle and plot the diagonal. Carefully measuring the length of the diagonal (the E_T vector) provides the amplitude of the total voltage. Using a protractor to measure angle θ gives you the amount of phase shift. If $E_R = 4$ V and $E_L = 3$ V, you will find that E_T is 5 V and that angle θ is about 37°.

**Figure 13-4.
Current and Voltage Vector Diagram for Series RL Circuits**

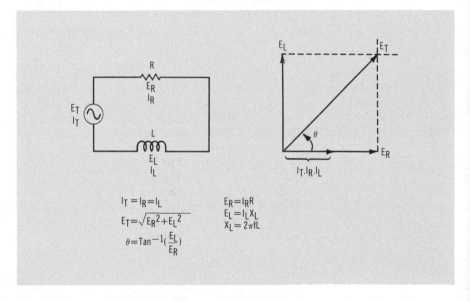

194   UNDERSTANDING ELECTRICITY AND ELECTRONICS CIRCUITS

UNDERSTANDING RL CIRCUITS

If you are prepared to handle the mathematics (preferably with the help of a scientific or engineering calculator), you can determine the amplitude of E_T and the phase angle far more accurately. Use the following equation to get the amplitude of the total voltage.

$$E_T = \sqrt{E_R^2 + E_L^2}$$

Suppose that the voltage across the resistance is 4 V and the voltage across the inductance is 3 V.

$$E_T = \sqrt{E_R^2 + E_L^2}$$
$$E_T = \sqrt{4^2 + 3^2}$$
$$E_T = \sqrt{16 + 9}$$
$$E_T = \sqrt{25}$$
$$E_T = 5 \text{ V}$$

Calculating the value of the phase angle, Θ, is a matter of finding an angle whose tangent is equal to E_L/E_R. This is done by calculating the Arctangent of that ratio. Arctangent functions on calculators are sometimes shown as Atan and other times as Tan^{-1}. We will use the latter notation. So the equation for determining the phase angle in a series RL circuit is:

$$\Theta = \text{Tan}^{-1}(E_L/E_R)$$

If $E_L = 3$ V and $E_R = 4$ V, as in the last example, the phase angle can be found this way:

$$\Theta = \text{Tan}^{-1}(E_L/E_R)$$
$$\Theta = \text{Tan}^{-1}(3/4)$$
$$\Theta = \text{Tan}^{-1}(0.75)$$
$$\Theta = 36.9°$$

Careful study of *Figure 13-4* yields two important characteristics of Series RL circuits.

1. The total voltage is greater than either the resistive or inductive voltage but less than the sum of the two.
2. The total voltage leads the total current by an angle less than 90°.

Total Opposition to Current Flow in a Series RL Circuit

We have seen that the total voltage in a series RL circuit is greater than either the voltage across the resistance or the voltage across the inductance, but it is less than the sum of those two voltages. The same general principle applies to the total opposition to current flow. The total

13 UNDERSTANDING RL CIRCUITS

opposition to current flow in a series RL circuit is greater than the amount of resistance and greater than the amount of inductive reactance, but it is less than the sum of resistance and inductive reactance.

You have already learned that you can determine the value of resistance by Ohm's law: $R = E_R/I_R$. That is, the value of a resistance is equal to the voltage across the resistance divided by the amount of current flowing through it. You have also learned that the inductive reactance of a coil is equal to the voltage across the inductance divided by the current flowing through it: $X_L = E_L/I_L$.

In a series RL circuit, the current has the same amplitude and phase through all components. The current through the resistance is the same as the current through the inductance, and both are the same as the circuit's total current. The voltages, however, are not in phase. As a result, the total opposition to current in a series RL circuit is not the simple sum of the values of resistance and inductive reactance.

Impedance is the combined opposition to current flow in a circuit that contains both resistance and reactance.

The overall opposition to current flow in a circuit that includes both resistance and reactance is something quite different from either of those qualities. This overall opposition to current flow cannot be called resistance because the current and voltage for a resistance are always in phase. It cannot be called a reactance because the current and voltage for a reactance are always 90° out of phase. Therefore, the overall opposition to current flow in such circuits must account for the fact that its current and voltage are out of phase, but by some angle less than 90 degrees. This quality of a circuit is called "impedance."

The basic unit of measurement for impedance is the ohm.

Impedance is the opposition to current flow in circuits that include resistance as well as reactance. The mathematical symbol for impedance is Z, and the basic unit of measurement is the ohm.

Figure 13-5 shows an impedance vector diagram for a series RL circuit. Notice that it looks very much like a voltage vector diagram for the same kind of circuit. The impedance diagram, however, shows the value of resistance as the horizontal vector and inductive reactance as the vertical vector. The diagonal of the rectangle formed by the resistance and reactance vectors is the impedance vector, Z. The length of the impedance vector specifies the value of impedance, and the angle of the impedance vector with respect to the resistance vector represents the phase angle.

**Figure 13-5.
Impedance Vector Diagram for Series RL Circuits**

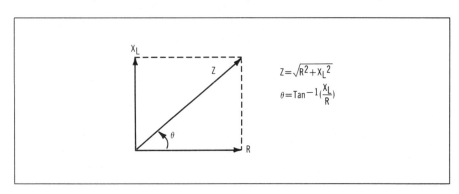

UNDERSTANDING RL CIRCUITS

If you are not familiar with calculating square roots and trigonometric functions, you can estimate the value of impedance by using the graphical methods described earlier for the voltage vector diagrams. Otherwise you can use this equation to determine the impedance of a series RL circuit:

$$Z = \sqrt{R^2 + X_L^2}$$

where,

Z is the impedance of the circuit in ohms,
R is the resistance in ohms,
X_L is the inductive reactance in ohms.

The phase angle shown in an impedance vector diagram for a series RL circuit indicates the phase shift between the total voltage and current in the same circuit.

The phase angle indicated on the impedance vector diagram for a series RL circuit has little meaning with regard to the values of resistance and reactance, but it happens to be the same phase angle you would find for a voltage diagram for the same circuit. The equation for the phase angle in this instance is:

$$\Theta = \operatorname{Tan}^{-1}(X_L/R)$$

where,

Θ is the phase angle in degrees,
X_L is the inductive reactance,
R is the resistance.

Suppose you have an RL circuit composed of a 200-Ω resistor connected in series with an inductor that has a reactance of 500 Ω. What is the impedance of the circuit and the phase angle?

From the information stated, you can see that R = 200 Ω and X_L = 500 Ω. Using the equation for impedance in a series RL circuit:

$$Z = \sqrt{R^2 + X_L^2}$$
$$Z = \sqrt{200^2 + 500^2}$$
$$Z = \sqrt{40{,}000 + 250{,}000}$$
$$Z = \sqrt{290{,}000}$$
$$Z = 539 \; \Omega$$

Notice that the impedance is indeed greater than the resistance or reactance but far less than the simple sum of the two.

As far as the phase angle is concerned:

$$\Theta = \operatorname{Tan}^{-1}(X_L/R)$$
$$\Theta = \operatorname{Tan}^{-1}(200/500)$$
$$\Theta = \operatorname{Tan}^{-1}(0.4)$$
$$\Theta = 21.8°$$

13 UNDERSTANDING RL CIRCUITS

Total Current in a Series RL Circuit

We know that the same current—both in amplitude and phase—flows through every part of a series RL circuit. Many practical situations do not directly indicate the current through such a circuit, so we must determine the series current by some other means.

In spite of the fact that the total voltage applied to a series RL circuit leads the total current, you can use Ohm's law to determine the total current if you know the total voltage and impedance.

$$I_T = \frac{E_T}{Z}$$

where,

I_T is the total current in amperes,
E_T is the total voltage in volts,
Z is the impedance of the circuit in ohms.

So if you apply 120 V ac to a circuit that has an impedance of 500 Ω, you can bet that the total current is 120/500, or 0.24 A.

**Figure 13-6.
Analysis of Series RL Circuit**

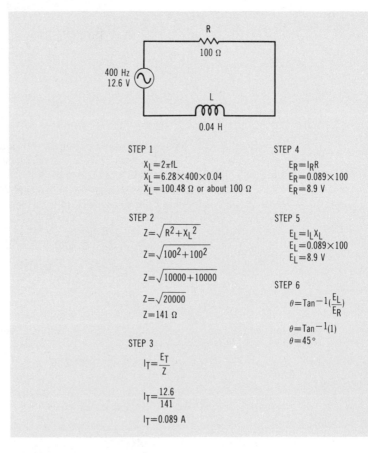

STEP 1
$X_L = 2\pi fL$
$X_L = 6.28 \times 400 \times 0.04$
$X_L = 100.48\ \Omega$ or about 100 Ω

STEP 2
$Z = \sqrt{R^2 + X_L^2}$
$Z = \sqrt{100^2 + 100^2}$
$Z = \sqrt{10000 + 10000}$
$Z = \sqrt{20000}$
$Z = 141\ \Omega$

STEP 3
$I_T = \frac{E_T}{Z}$
$I_T = \frac{12.6}{141}$
$I_T = 0.089$ A

STEP 4
$E_R = I_R R$
$E_R = 0.089 \times 100$
$E_R = 8.9$ V

STEP 5
$E_L = I_L X_L$
$E_L = 0.089 \times 100$
$E_L = 8.9$ V

STEP 6
$\theta = \tan^{-1}\left(\frac{E_L}{E_R}\right)$
$\theta = \tan^{-1}(1)$
$\theta = 45°$

Analyzing Series RL Circuits

You now have enough information at hand to do a complete analysis of a series RL circuit. *Figure 13-6* shows a circuit that includes the amplitude and frequency of the ac voltage source, the value of a resistor, and the value of inductance of a coil. A complete analysis of this circuit would include finding the amount of current flowing through the circuit, the reactance of the coil, the impedance of the circuit, the voltage drop across each component, and the phase angle between the source voltage and current.

Step 1 in the procedure takes advantage of the only calculation that can be done from the given information—finding the reactance of the inductor. Knowing the reactance of the inductor then lets you calculate the impedance of the circuit, as shown in Step 2. The procedure in Step 3 uses the impedance and the given source voltage to find the current through the circuit.

Knowing the current lets you calculate the voltage across the resistor (Step 4) and the voltage across the inductor (Step 5). Finally, you are in a position to determine the phase angle as shown in Step 6.

The results of this analysis can be summarized as shown in *Table 13-1*.

Table 13-1. Summary of Analysis in Figure 13-6

Circuit		Resistor	
Total voltage: E_T = 12.6 V	(Given)	Resistance value: R = 100 Ω	(Given)
Frequency: f = 400 Hz	(Given)	Resistor voltage: E_R = 8.9 V	(Step 4)
Total current: I_T = 0.089 A	(Step 3)	Resistor current: I_R = 0.089 A	(From I_T; series circuit)
Impedance: Z = 141 Ω	(Step 2)		
Phase shift: Θ = 45°	(Step 6)		
Inductor			
Inductance Value: L = 0.04 H	(Given)		
Inductive reactance: X_L = 100 Ω	(Step 1)		
Inductor voltage: E_L = 8.9 V	(Step 5)		
Inductor current: I_L = 0.089 A	(From I_T; series circuit)		

UNDERSTANDING RL CIRCUITS

WHAT IS THE Q OF AN INDUCTOR?

We have stated a number of times that there is no such thing as a perfect inductor. Inductors are constructed from turns of wire and although wire is usually a good conductor, it always possesses some resistance to current flow. In other words, a practical inductor has a bit of resistance as well as some amount of inductance. The qualities of inductance and resistance in a coil of wire can be represented as a series RL circuit. A practical inductor, in fact, exhibits all the qualities of a series RL circuit.

Figure 13-7 illustrates a practical inductor. The inductor is rated at 0.1 H, but the wire that makes up its coil has 10-Ω resistance. An ideal inductor would have a resistance of 0 Ω. The larger the resistance of the coil, compared to its reactance, the less ideal the inductor becomes.

A practical inductor includes some resistance in its coil of wire, so a practical inductor can be portrayed as a pure inductance connected in series with a resistance.

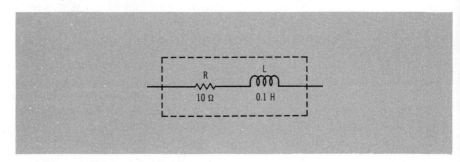

Figure 13-7. Practical Inductors Include Some Resistance

A measure of the quality of an inductor is its "Q factor." The Q of a coil is expressed in this fashion:

$$Q = \frac{X_L}{R}$$

where,

The Q factor of an inductor is an indication of its quality.

Q is the Q factor (unitless),
X_L is the inductive reactance of the coil in ohms,
R is the resistance of the coil in ohms.

You can see that smaller values of R provide higher values of Q. Bear in mind that an ideal inductor would have no resistance and the inductor would have an infinite Q.

The value of R does not change with frequency, but the value of X_L increases in proportion to the operating frequency. You can see that increasing the frequency applied to an inductor increases its Q factor by a proportional amount.

The Q factor of an inductor changes in proportion to the operating frequency.

PARALLEL RL CIRCUITS

Parallel RL circuits share at least one important characteristic with any other kind of parallel circuit; the voltage is the same across every component in the parallel arrangement. You will find, however, that total

UNDERSTANDING RL CIRCUITS

current is less than the sum of the currents through the individual branches.

Voltage in a Parallel RL Circuit

The amplitude and phase of voltages are identical in all parts of a parallel RL circuit.

The same voltage is dropped across every component in a parallel circuit. This means the amplitude of the voltage across every component is equal to the amplitude of the total voltage:

$$E_T = E_L = E_R$$

where,

E_T is the total voltage in volts,
E_L is the voltage across the inductance in volts,
E_R is the voltage across the resistance in volts.

The fact that the same voltage appears across every component in a parallel circuit also means that the phase of the voltage is the same for each component.

Figure 13-8 illustrates these important ideas about voltage across the components in a parallel RL circuit. The sine-wave diagrams show that the voltages across the resistance and inductance have the same amplitude and phase as the total voltage (the source voltage in this instance). The voltage vector diagram illustrates the same principle: the voltage vectors all have the same length and direction, so they appear superimposed on one another.

**Figure 13-8.
Voltages in Parallel RL Circuits**

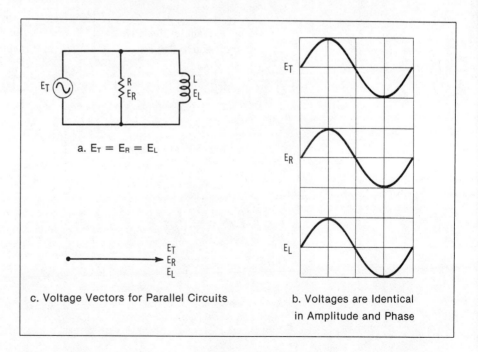

a. $E_T = E_R = E_L$

c. Voltage Vectors for Parallel Circuits

b. Voltages are Identical in Amplitude and Phase

13 UNDERSTANDING RL CIRCUITS

Current in a Parallel RL Circuit

The amount of current flowing through a resistance is equal to the amplitude of the voltage divided by the value of the resistance: $I_R = E_R/R$. Furthermore, the voltage and current waveforms are in phase, as shown in the resistor waveforms in *Figure 13-9a*.

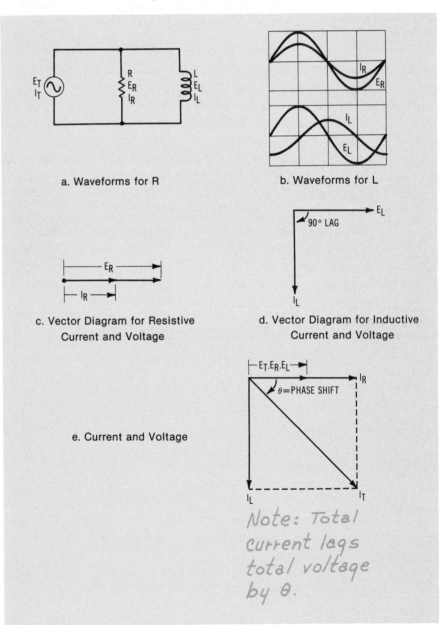

**Figure 13-9.
Currents and Voltages in Parallel RL Circuits**

a. Waveforms for R

b. Waveforms for L

c. Vector Diagram for Resistive Current and Voltage

d. Vector Diagram for Inductive Current and Voltage

e. Current and Voltage

Note: Total current lags total voltage by θ.

UNDERSTANDING RL CIRCUITS

The current flowing through an inductance is equal to the amplitude of the voltage dropped across it divided by the value of inductive reactance: $I_L = E_L/X_L$. The current through the inductance, however, lags its voltage drop by exactly 90°.

Now compare the waveforms for the resistance and inductance for a parallel RL circuit. You can see that the voltages are in phase; this is a fixed principle for any parallel circuit. The currents, however, are 90° out of phase. The current through the inductance lags the voltage across the resistance by 90°.

The current through the inductor in a parallel RL circuit lags the voltage across the resistance by 90°.

The vector diagrams, *Figure 13-9c* and *d*, for the resistance and inductance show the phase relationships for those components. The vector diagram for the resistance shows the voltage and current vectors pointing in the same direction. The vector diagram for the inductance shows the inductive current lagging the inductive voltage by 90°.

Because the amplitude and phase of the voltages across the resistance and inductance are the same in a parallel circuit, you can combine their vector diagrams by aligning their voltage vectors. The result is a vector diagram for the current and voltage as shown in *Figure 13-9e*. This diagram shows the voltages having the same amplitude and phase, the current through the resistor having the same phase as the voltages, and the current through the inductance lagging by 90°.

The total current in a parallel RL circuit is not equal to the sum of the resistive and inductive currents.

The fact that the inductive and resistive currents are 90° out of phase means that the total current in a parallel RL circuit is *not* equal to the sum of the resistive and inductive currents. The final diagram in *Figure 13-9* shows that the total current vector is represented by the diagonal of a rectangle, where the sides of the rectangle are the resistive and inductive currents. The length of the I_T vector indicates the amount of total current, and the angle between the I_T and I_R vectors represents the phase angle.

Figure 13-10 is a vector diagram for parallel RL circuits. You can determine the amplitude of the total current and the phase angle by means of the graphical methods described previously for a series RL circuit. But if you are prepared to handle the mathematics, you can determine the amplitude of I_T and the phase angle with the help of the following equations:

$$I_T = \sqrt{I_R^2 + I_L^2}$$

and

$$\Theta = \text{Tan}^{-1}(I_L/I_R)$$

**Figure 13-10.
Currents and Voltage
Vector Diagram for
Parallel RL Circuits**

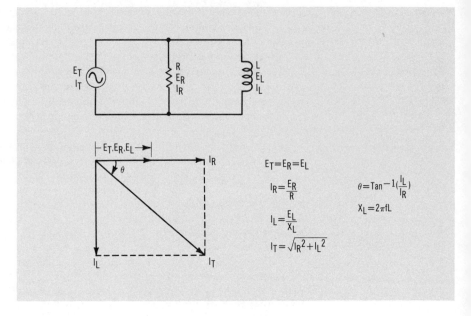

Suppose that the current through the resistance is 3 A and the current through the inductance is 4 A. Using the equation for the total current in a parallel RL circuit:

$$I_T = \sqrt{I_R^2 + I_L^2}$$
$$I_T = \sqrt{3^2 + 4^2}$$
$$I_T = \sqrt{9 + 16}$$
$$I_T = \sqrt{25}$$
$$I_T = 5 \text{ A}$$

Calculating the phase angle in a parallel RL circuit:

$$\theta = \text{Tan}^{-1}(I_L/I_R)$$
$$\theta = \text{Tan}^{-1}(4/3)$$
$$\theta = \text{Tan}^{-1}(1.33)$$
$$\theta = 53.1°$$

Careful study of the current/voltage vector diagram in *Figure 13-10* yields two characteristics of parallel RL circuits.

1. Total current is greater than either the resistive or inductive currents, but less than the sum of the two.
2. Total current lags the total voltage by some angle less than 90°.

UNDERSTANDING RL CIRCUITS

Impedance of a Parallel RL Circuit

Recall that the impedance of a circuit is the combination of opposition to current flow from the circuit's resistive and inductive components. Regarding the impedance of series RL circuits, you have seen that you can use impedance vector diagrams, square-root equations, and Arctangent functions to work with the impedance and phase angle. Matters are not quite so simple for the impedance in parallel RL circuits. In fact, the topic calls for more discussion than we want here.

We need to use a more roundabout procedure for determining the impedance of a parallel RL circuit. The most satisfactory procedure for our purposes is to use this form of Ohm's law:

$$Z = \frac{E_T}{I_T}$$

where,

Z is the impedance of the circuit in ohms,
E_T is the total voltage in volts,
I_T is the total current in amperes.

You can calculate the impedance of a parallel RL circuit on the basis of values for total voltage that you make up for yourself.

Suppose you have an RL circuit composed of a 200-Ω resistor connected in parallel with an inductor that has a reactance of 500 Ω. What is the impedance of the circuit? You cannot directly answer this question because the equation for impedance of a parallel RL circuit is expressed in terms of total voltage and current. You can handle the situation, though, by making up any convenient value for source voltage. Having a source voltage lets you apply Ohm's law to calculate the currents through the resistance and inductance. After doing that, you can use the formula for total current in a parallel RL circuit to find a total current. And finally, you can use the equation just cited to calculate the impedance. Let's walk through that procedure, step by step.

1. You are given $R = 200\ \Omega$ and $X_L = 500\ \Omega$.
2. Pick a convenient voltage such as 100 V, and use it to calculate the currents through the components:

$$I_R = E_R/R$$
$$I_R = 100/200$$
$$I_R = 0.5\ \text{A}$$
$$I_L = E_L/X_L$$
$$I_L = 100/500$$
$$I_L = 0.2\ \text{A}$$

3. Use those values of current to calculate the total current in a parallel RL circuit:

$$I_T = \sqrt{I_R^2 + I_L^2}$$
$$I_T = \sqrt{0.5^2 + 0.2^2}$$
$$I_T = \sqrt{0.25 + 0.04}$$
$$I_T = \sqrt{0.29}$$
$$I_T = 0.539 \text{ A}$$

4. Use your value of total current and voltage to calculate the impedance of the circuit:

$$Z = E_T/I_T$$
$$Z = 100/0.539$$
$$Z = 186 \text{ }\Omega$$

Analyzing Parallel RL Circuits

You now have enough information at hand to do a complete analysis of a parallel RL circuit. *Figure 13-11* shows a circuit that includes the amplitude and frequency of the ac voltage source, the value of a resistor, and the value of the inductance of a coil. A complete analysis of this circuit includes the voltage drop across each component in the circuit, the reactance of the coil, the impedance of the circuit, the current through each component, the total current, and the phase angle between the source voltage and current. The results are summarized for you in *Table 13-2*.

HOW DO RL CIRCUITS AFFECT PULSE WAVEFORMS?

Most of the features of RL circuits described so far assume that the components are operating from an ac, sine-wave source. This is indeed the most common waveform used with series and parallel RL circuits. However, there are also many instances where pulse waveforms are applied to RL circuits.

UNDERSTANDING RL CIRCUITS

**Figure 13-11.
Analysis of Parallel RL Circuit**

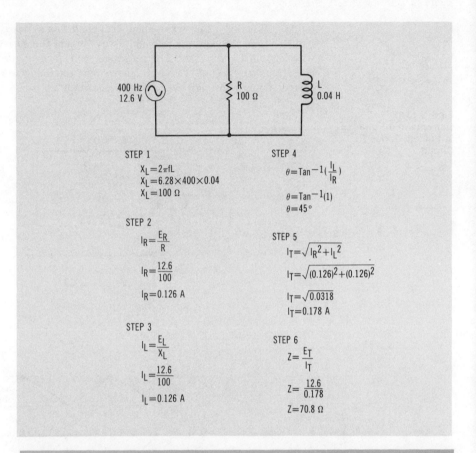

STEP 1
$X_L = 2\pi fL$
$X_L = 6.28 \times 400 \times 0.04$
$X_L = 100\ \Omega$

STEP 2
$I_R = \dfrac{E_R}{R}$

$I_R = \dfrac{12.6}{100}$

$I_R = 0.126\ A$

STEP 3
$I_L = \dfrac{E_L}{X_L}$

$I_L = \dfrac{12.6}{100}$

$I_L = 0.126\ A$

STEP 4
$\theta = \text{Tan}^{-1}\left(\dfrac{I_L}{I_R}\right)$

$\theta = \text{Tan}^{-1}(1)$
$\theta = 45°$

STEP 5
$I_T = \sqrt{I_R^2 + I_L^2}$

$I_T = \sqrt{(0.126)^2 + (0.126)^2}$

$I_T = \sqrt{0.0318}$
$I_T = 0.178\ A$

STEP 6
$Z = \dfrac{E_T}{I_T}$

$Z = \dfrac{12.6}{0.178}$

$Z = 70.8\ \Omega$

**Table 13-2.
Summary of Analysis in Figure 13-11**

Circuit		Resistor	
Total voltage: E_T = 12.6 V	(Given)	Resistance value: R = 100 Ω	(Given)
Frequency: f = 400 Hz	(Given)	Resistor voltage: E_R = 12.6 V	(From E_T; parallel circuit)
Total current: I_T = 0.178 A	(Step 5)	Resistor current: I_R = 0.126 A	(Step 2)
Impedance: Z = 70.8 Ω	(Step 6)		
Phase shift: Θ = 45°	(Step 4)		
Inductor			
Inductance value: L = 0.04 H (Given)			
Inductive reactance: X_C = 100 Ω (Step 1)			
Inductor voltage: E_C = 12.6 V (From E_T; parallel circuit)			
Inductor current: I_C = 0.126 A (Step 3)			

13 UNDERSTANDING RL CIRCUITS

When a square-wave pulse is applied to a series RL circuit, the inductor tends to slow the rise and decay time of the current. The first waveform in *Figure 13-12* shows a square waveform applied to a series RL circuit, and the second waveform shows how the inductance affects the current flowing through every part of the circuit.

**Figure 13-12.
Currents and Voltages
in Pulsed, Series RL
Circuit**

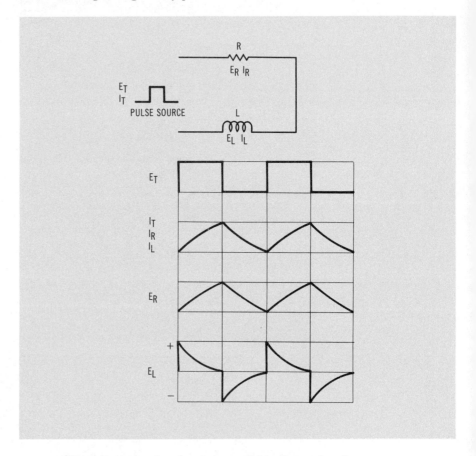

The voltage developed across a resistor always has the same waveform as its current. So you can see that the voltage across the resistor in a pulsed, series RL circuit has the same rounded-off shape as the current.

The voltage developed across the inductor in this circuit has a waveform that is altogether different from the others. Notice how the inductor voltage is peaked with every change in the voltage from the source. The positive-going edge of the applied waveform creates positive peaks across the inductor. The negative-going edge of the applied waveform produces negative peaks of voltage across the inductor.

UNDERSTANDING RL CIRCUITS

Pulsed RL circuits are described in terms of their time constant. For a series RL circuit, the time constant is defined as the time required for changing the current by 63%. As indicated in *Figure 13-13*, the time constant of an RL circuit is equal to the value of the inductor divided by the value of the resistor:

$$T = \frac{L}{R}$$

where,

T is the time constant in seconds,
L in the inductance in henrys,
R is the resistance in ohms.

**Figure 13-13.
Time Constant for
Series RL Circuit**

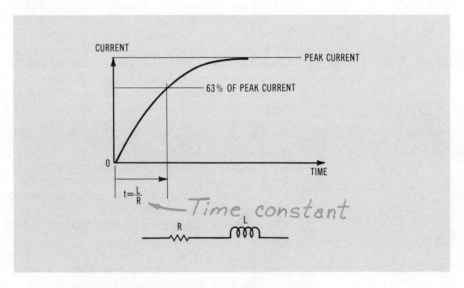

Suppose that you have a series circuit composed of a 100-Ω resistor and a 0.01-H inductor. What is the time constant of this circuit? From the equation for the time constant of a series RL circuit:

$$T = L/R$$
$$T = 0.01/100$$
$$T = 0.0001 \text{ second}$$

WHAT HAVE WE LEARNED?

1. An RL circuit is one that includes both a resistance and an inductance.
2. Currents are equal in amplitude and phase through all parts of a series RL circuit.

3. Current and voltage are in phase through the resistance in a series RL circuit.
4. Voltage leads the current by 90° for the inductance in a series RL circuit.
5. The total current lags the total voltage for a series RL circuit. The amount of current lag is less than 90°.
6. The total voltage in a series RL circuit is greater than the resistive or inductive voltages but less than the sum of the two.
7. You can represent the amplitudes and phase relationships in a series RL circuit by means of vector diagrams.
8. You can use graphical methods to determine the total voltage and phase angle in a series RL circuit.
9. Impedance is the opposition to current flow in ac circuits that contain both resistance and reactance. The mathematical symbol for impedance is Z, and the unit of measurement is the ohm.
10. The impedance of a series RL circuit is greater than either the resistance or inductive reactance but less than the sum of the two.
11. An impedance vector diagram for a series RL circuit shows the relationships between resistance, inductive reactance, impedance, and the phase angle.
12. Every inductor includes a bit of resistance in its windings. The quality, or Q factor, of an inductor is the ratio of inductive reactance to winding resistance. The Q factor for an inductor is a unitless term.
13. Voltages are equal in amplitude and phase across all parts of a parallel RL circuit.
14. Current lags the voltage by 90° for the inductance in a parallel RL circuit.
15. The total current lags the total voltage for a parallel RL circuit. The amount of current lag is less than 90°.
16. The total current in a parallel RL circuit is greater than the resistive or inductive currents but less than the sum of the two.
17. You can represent the amplitudes and phase relationships in a parallel RL circuit by means of vector diagrams.
18. You can use graphical methods to determine the total current and phase angle in a parallel RL circuit.
19. There is no such thing as an impedance vector diagram for parallel RL circuits. You can calculate the impedance of a parallel RL circuit, however, by dividing the total voltage by the total current.
20. The inductor in a series RL circuit rounds the edges of square-wave currents and peaks the edges of the voltage across the inductor.
21. The time constant of a circuit is the time required for the waveform to change 63% in amplitude.
22. The time constant of a series RL circuit is equal to the inductance divided by the resistance. The result is in units of seconds.

KEY WORDS

Impedance
Time constant

Quiz for Chapter 13

1. Which one of the following statements is true for the voltages in a series RL circuit?
   a. The voltage always has the same amplitude and phase for every part of the circuit.
   b. The total voltage is equal to the sum of the voltages across the resistance and inductance.
   c. The total voltage is less than the sum of the voltages across the resistance and inductance.
   d. The total voltage lags the total current by less than 90°.

2. Which one of the following statements is true for the currents in a series RL circuit?
   a. The current always has the same amplitude and phase for every part of the circuit.
   b. The total current is equal to the sum of the currents through the resistance and inductance.
   c. The total current is less than the sum of the currents through resistance and inductance.
   d. The total current leads the total voltage by less than 90°.

3. Which one of the following statements is true for the impedance of a series RL circuit?
   a. The impedance is equal to the sum of the resistance and inductive reactance.
   b. The impedance is greater than the sum of the resistance and inductive reactance.
   c. The impedance is greater than either the resistance or inductive reactance, but less than their sum.
   d. The impedance is less than the sum of the resistance and inductive reactance.
   e. The impedance cannot be calculated directly.

4. Which one of the following statements is true for the voltage in a parallel RL circuit?
   a. The voltage always has the same amplitude and phase for every part of the circuit.
   b. The total voltage is equal to the sum of the voltages across the resistance and inductance.
   c. The total voltage is less than the sum of the voltages across the resistance and inductance.
   d. The total voltage lags the total current by less than 90°.

5. Which one of the following statements is true for the currents in a parallel RL circuit?
   a. The current always has the same amplitude and phase for every part of the circuit.
   b. The total current is equal to the sum of the currents through the resistance and inductance.
   c. The total current is less than the sum of the currents through the resistance and inductance.
   d. The total current leads the total voltage by less than 90°.

6. Referring to the vector diagrams in *Figure 13-14*, which set best represents the distribution of voltages in a series RL circuit?
   a. Vectors 1A, 1B, and 1C.
   b. Vectors 2A, 2B, and 2C.
   c. Vector 1A only.
   d. Vector 2A only.

**Figure 13-14.
Diagrams for Questions
6–11**

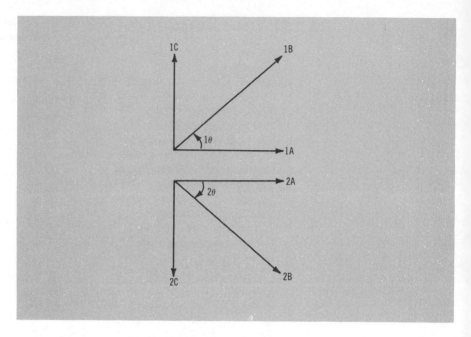

7. Referring to the vector diagrams in *Figure 13-14*, which set best represents the distribution of currents in a parallel RL circuit?
   a. Vectors 1A, 1B, and 1C.
   b. Vectors 2A, 2B, and 2C.
   c. Vector 1A only.
   d. Vector 2A only.

8. Referring to the vector diagrams in *Figure 13-14*, which set best represents an impedance diagram for a series RL circuit?
   a. The set made up of vectors 1A, 1B, and 1C.
   b. The set made up of vectors 2A, 2B, and 2C.
   c. Series impedance cannot be represented by a vector diagram.

9. Referring to the vector diagrams in *Figure 13-14*, which set best represents an impedance diagram for a parallel RL circuit?
   a. The set made up of vectors 1A, 1B, and 1C.
   b. The set made up of vectors 2A, 2B, and 2C.
   c. Parallel impedance cannot be represented by a vector diagram.

10. Referring to the vector diagrams in *Figure 13-14*, which vector best represents the total voltage in a series RL circuit?
    a. Vector 1A.
    b. Vector 1B.
    c. Vector 1C.
    d. Vector 2A.
    e. Vector 2B.
    f. Vector 2C.

UNDERSTANDING RL CIRCUITS

11. Referring to the vector diagrams in *Figure 13-14*, which vector best represents the total current in a parallel RL circuit?
    a. Vector 1A.
    b. Vector 1B.
    c. Vector 1C.
    d. Vector 2A.
    e. Vector 2B.
    f. Vector 2C.

12. On a vector diagram, leading phase angles are represented by:
    a. clockwise angles.
    b. counterclockwise angles.

13. Which one of the following statements is true for the Q factor of an inductor?
    a. The Q of an inductor increases as the applied frequency increases.
    b. The Q of an inductor decreases as the applied frequency increases.
    c. There is no relationship between the Q of an inductor and the applied frequency.

14. What is the impedance of the circuit in *Figure 13-15*?
    a. 19.3 Ω.
    b. 40.7 Ω.
    c. 50.1 Ω.
    d. 59 Ω.
    e. 81.4 Ω.

15. What is the phase angle between total current and voltage for the the circuit in Figure 13-15?
    a. 0°.
    b. 32.1°.
    c. 57.9°.
    d. 90°.

16. What is the total current for the circuit in *Figure 13-15*?
    a. 1.23 A.
    b. 1.69 A.
    c. 2 A.
    d. 2.46 A.
    e. 5.8 A.

17. What is the voltage across the resistor in Figure 13-15?
    a. 46.9 V.
    b. 53.1 V.
    c. 61.5 V.
    d. 84.5 V.
    e. 100 V.
    f. 123 V.

Figure 13-15.
Circuit for Questions 14–17

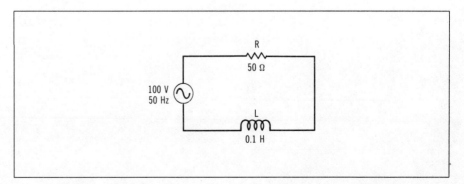

UNDERSTANDING ELECTRICITY AND ELECTRONICS CIRCUITS

18. What is the impedance of a circuit composed of a 100-Ω resistor connected in parallel with an inductor that has a reactance of 200 Ω?
   a. 66.7 Ω.
   b. 89.3 Ω.
   c. 150 Ω.
   d. 224 Ω.
   e. 300 Ω.
   f. Cannot be determined from the information supplied.

19. What is the phase shift between total current and voltage in the circuit described in Question 18?
   a. 0°.
   b. 26.6°.
   c. 45°.
   d. 63.4°.
   e. 90°.
   f. Cannot be determined from the information supplied.

Understanding Capacitance

ABOUT THIS CHAPTER

We have already studied two important electronic components, resistors and inductors. In this chapter, we examine a third kind of component—the capacitor. You will see how a capacitor stores electrical charges and how it affects the operation of ac circuits. You will become familiar with the units of measurement for capacitance and learn to calculate capacitive reactance.

CAPACITANCE

A capacitor is made up of two conductors separated by a dielectric (insulating) material.

The most fundamental definition of a capacitor is two conductors separated by an insulating material called a "dielectric." *Figure 14-1* illustrates the basic idea of a capacitor. There are two rectangular metal plates separated by air. The metal plates serve as the two conductors, and the air between them is the dielectric.

Before describing how the capacitor works, it is important to recall a few principles described in the first chapters of this book. First, you should remember that electrical current is defined as the movement of free electrons through a conductor. All conductors, including most metals, require very little electromotive force (voltage) to free those electrons and cause them to flow from a point of excess electrons to one having a deficiency of electrons. In a complete circuit, this means that electrons leave from the negative terminal of the voltage source, pass through the circuit, and return to the positive terminal of the voltage source.

Current does not flow between the two conductors of a capacitor because of the dielectric material that separates them.

The capacitor circuit, however, does not appear to be a complete circuit. There is an air gap between the two plates, and air, being a fairly good insulator, cannot allow electrons to complete the circuit from the negative terminal of the battery and back to the positive terminal. Indeed, current cannot flow between the plates (through the insulating material) of a capacitor.

14 UNDERSTANDING CAPACITANCE

**Figure 14-1.
Capacitor Principles**

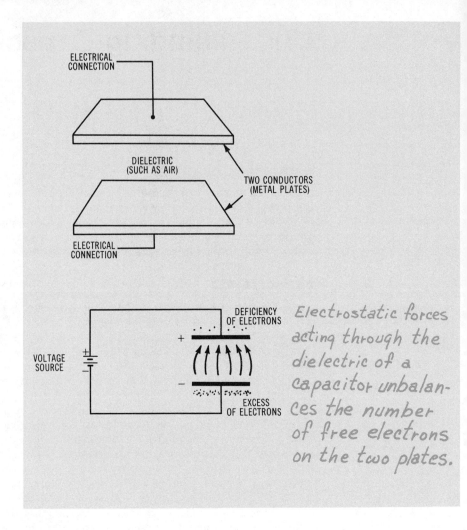

Electrostatic forces acting through the dielectric of a capacitor unbalances the number of free electrons on the two plates.

UNDERSTANDING CAPACITANCE

It would thus seem that a capacitor is simply an open-circuit device, much like a switch that is always open. That is not the case, however. The explanation can be found by recalling that electrons are surrounded by an electrostatic field that can exert a force on other charged particles without actually touching them. The larger the concentration of electrons, the stronger their electrostatic field becomes.

The electrostatic field between the plates of a capacitor creates a force that allows an excess of electrons to build up on the negative plate and pushes electrons off the positive plate.

By connecting the negative terminal of a voltage source to one plate of a capacitor and the positive terminal of the battery to the second plate, the concentration of electrons from the negative terminal of the battery creates an electrostatic field that extends through the insulating material. This negatively charged electrostatic field repels electrons from the positive plate. The plates thus become electrically charged. The plate connected to the negative terminal of the battery takes on an excess of electrons, and the plate connected to the positive terminal has a deficiency of electrons.

Current flows through a capacitor circuit (but not through the capacitor itself) until the voltage on the plates of the capacitor is equal to that of the voltage source.

Figure 14-2a shows a fundamental capacitor that has no charge on its plates. When a voltage source is first applied to the plates of the capacitor (as shown *Figure 14-2b*), electrons flow from the negative terminal of the battery to the plate that is connected to that terminal. At the same time, an equal number of electrons flow from the positive plate to the positive terminal of the battery. This movement of electrons, or current flow, continues until the plates are charged to the same voltage as the voltage source. During that "charge time," current flows through the voltage source and connections to the capacitor, but never between the plates of the capacitor.

A capacitor can retain its charge after it is removed from its voltage source.

Now suppose that you have charged a capacitor. What happens if you carefully remove the capacitor from the voltage source? The charge actually remains on the plates of the capacitor. Ideally, the capacitor would remain charged forever. In actual practice, however, there is always a small amount of leakage of electrons between the plates and around the outer structure of the capacitor. Capacitors that have been carefully disconnected from a voltage source can hold their charge for several seconds up to several hours, depending on the value of capacitance and the quality of its manufacture.

14 UNDERSTANDING CAPACITANCE

Figure 14-2.
Charging a Capacitor

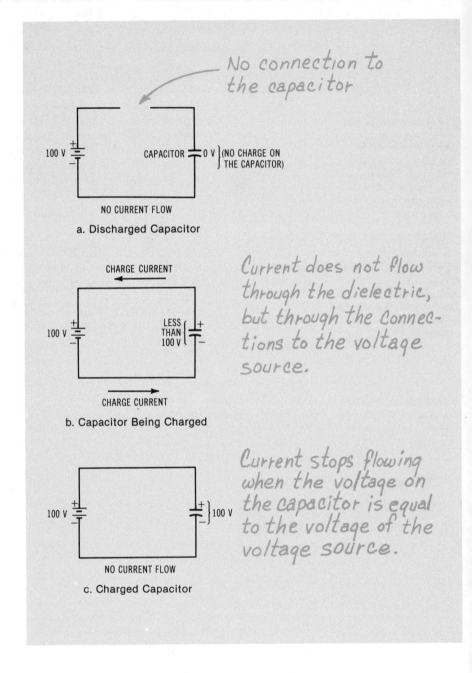

UNDERSTANDING CAPACITANCE

14

During the time a capacitor is being discharged, it acts as a source of voltage and current. A capacitor is completely discharged when there is no longer a difference in potential between its plates.

Figure 14-3 suggests one way to discharge a capacitor that has been charged from a voltage source. Connecting a wire between the two terminals discharges the capacitor rather quickly. During that "discharge time," electrons flow through the wire, from the negatively charged plate to the positive plate. This discharge current continues to flow until you remove the wire or until the capacitor is completely discharged (there is no difference in potential between the two plates). During the time a capacitor is being discharged, it actually acts as a source of voltage and current.

**Figure 14-3.
Discharging a Capacitor**

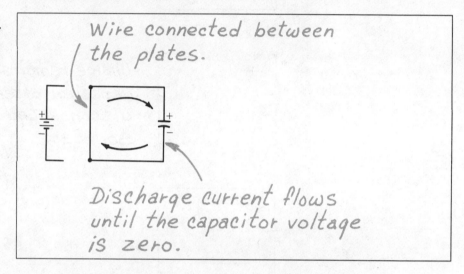

Figure 14-4 shows a different kind of capacitor charging-and-discharging situation. The first diagram shows a capacitor being charged from a voltage source. As usual, current flows through the circuit (but not between the plates of the capacitor) until the voltage on the capacitor is equal to the voltage of the source. The second diagram shows the charge capacitor disconnected from the voltage source. The energy that was required for charging the capacitor is stored as an electrostatic field. The third diagram shows the capacitor connected to the source, but the polarity of the source is reversed. This not only discharges the capacitor, but it subsequently recharges it with the new source polarity. A substantial amount of current can flow while the capacitor is discharging and then recharging with the opposite polarity.

The charge, discharge, and recharge current for a capacitor in an ac circuit allows ac current to flow back and forth through the circuit, even though the current never actually flows through the dielectric material in the capacitor.

This example of charging a capacitor at one polarity then discharging and recharging at the opposite polarity sheds some light on how capacitors behave in ac circuits. In ac circuits, the capacitor is continually being charged, discharged, and recharged with the opposite polarity. Even though there is never any current flow between the plates of the capacitor, itself, the ac circuit behaves as though there is a complete circuit for ac current flow.

UNDERSTANDING ELECTRICITY AND ELECTRONICS CIRCUITS   219

**Figure 14-4.
Changing the Voltage Polarity Applied to Capacitor**

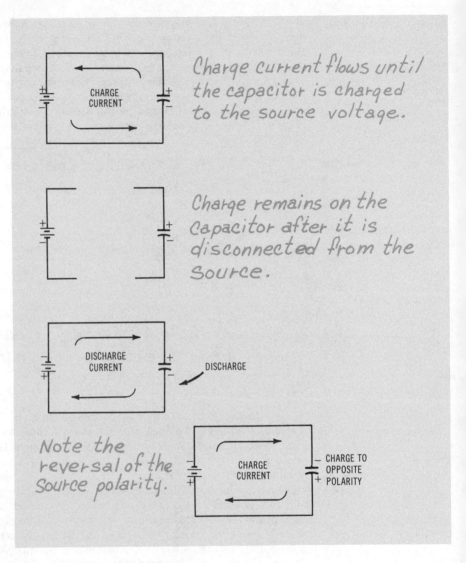

RATING CAPACITORS

Capacitors are rated according to their value of capacitance and maximum working voltage. The basic unit of measurement for capacitance is the farad (abbreviated F). A 1-F capacitor requires 1 A of current to change its charge at the rate of 1 V per second.

The unit of measurement for capacitance is the farad. A 1-farad capacitor requires 1 ampere of current to change its voltage at the rate of 1 volt per second.

UNDERSTANDING CAPACITANCE

14

Most capacitors are rated in microfarads or picofarads. One microfarad is one-millionth of a farad, and one picofarad is one-millionth of a microfarad. Exceeding the maximum operating voltage rating of a capacitor breaks down the dielectric and risks a hazardous explosion or fire.

In actual practice, you rarely find capacitance values that come close to 1 F. Most capacitors have much smaller values. Most capacitors are rated in microfarads (millionths of a farad) and picofarads (millionths of a microfarad). The abbreviation for a microfarad is μF, and a picofarad is abbreviated pF.

The maximum operating voltage of a capacitor is also an important consideration. No dielectric is a perfect insulator, and there is always a limit on how much voltage can be applied before the insulator breaks down and conducts current. If you exceed the voltage rating of a capacitor, you run the risk of breaking down its dielectric and destroying the capacitor. Exceeding the voltage rating of a capacitor can cause a potentially harmful explosion or fire.

Most capacitors have their capacitance value and maximum working voltages printed on them. Sometimes the values are abbreviated, but practical experience and common sense will help you read such abbreviations reliably.

CAPACITORS' CONSTRUCTION

Capacitors, like resistors, can be classified according to the materials used in manufacturing them or to the way they are constructed. But no matter what materials are used or how they are assembled, all capacitors are basically two conductors separated by a dielectric material.

The most common dielectric materials are air, mica, ceramics, Mylar, oil, and a paste-like electrolyte material. The smallest-valued capacitors, those in the picofarad range, are smaller than a dime and have a wafer or button shape. These have dielectrics made of mica or a ceramic material. Somewhat larger-valued capacitors, those having values from $0.001~\mu F$ to $1~\mu F$, generally use dielectrics made of Mylar or oil. Mylar-filled capacitors are used in electronic circuits, and oil-filled capacitors are generally used in high-voltage industrial machinery.

Always observe the polarity markings (+ and −) when connecting an electrolytic capacitor into a circuit.

Capacitors using an electrolyte as a dielectric material have the largest capacity. These electrolytic capacitors have values ranging from one to tens of thousands of microfarads. You must use special care when connecting electrolytic capacitors into circuits because electrolytic capacitors are polarized. That is, one terminal must always be connected to a positive source and the other to a negative source. The polarities, positive and negative, are clearly marked beside the connections on electrolytic capacitors. Connecting an electrolytic capacitor into a circuit with its + and − connections reversed will depolarize the electrolyte and quickly destroy the capacitor. Can you explain why an electrolytic capacitor should never be used in an ac circuit?

Figure 14-5 shows how some kinds of capacitors are constructed. These methods of construction allow a large area for the plates, but confine it to a relatively small space. Capacitors that are rolled into a rod shape usually have a stripe painted around one end. That stripe indicates the connection to the outer plate (or foil). In high-frequency circuits, such as those in radio and television equipment, you should always connect the capacitor so that the stripe is closer to the circuit's ground reference point.

UNDERSTANDING CAPACITANCE

**Figure 14-5.
Construction of Some Capacitors**

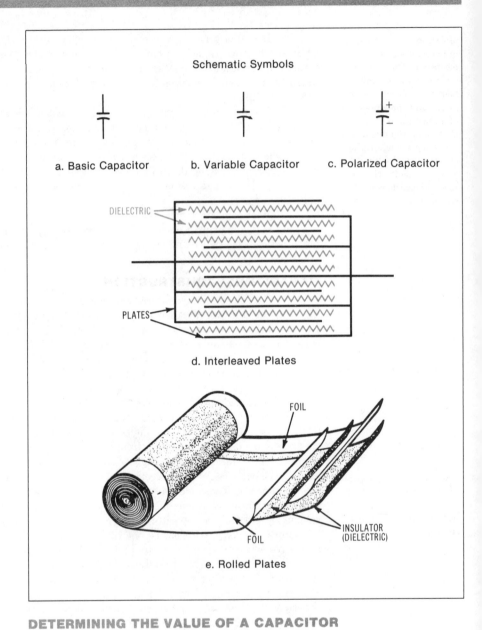

DETERMINING THE VALUE OF A CAPACITOR

Capacitance is proportional to the area of the plates; the larger the plate area, the larger the capacitance.

The amount of electrical charge that can be stored on a capacitor (the quantity of electrons that can be placed on the plates) varies with the area of those plates. Consequently, capacitance varies directly with area. When the area is doubled (by using larger plates or doubling the number of plates connected in parallel) the capacitance doubles.

UNDERSTANDING CAPACITANCE

Capacitance is inversely proportional to the distance between the plates. The smaller the distance, the greater the capacitance.

As indicated in *Figure 14-6*, capacitance is inversely proportional to the distance between the plates. The greater the distance between the plates, the smaller the amount of capacitance.

**Figure 14-6.
Factors Affecting Capacitor Values**

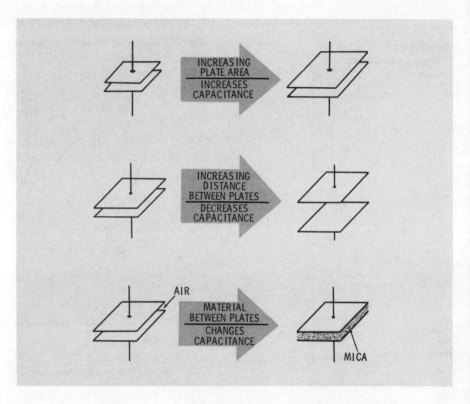

The dielectric strength of a material is the measure of its ability to withstand electric stress without breaking down.

The dielectric constant of a material is a measure of how well it concentrates electrostatic lines of force.

It is possible to obtain higher values of capacitance by using a dielectric other than air. This allows the plates to be placed closer together without letting the charge cross the gap. Dielectrics such as mica, glass, oil, and Mylar are a few of the materials that can sustain a high electric stress without breaking down. This property is called the "dielectric strength."

Besides allowing the plates to be brought closer together, the dielectric has another effect on capacitance. Some dielectrics can concentrate electrostatic lines of force better than others. This property is measured in terms of "dielectric constant." The larger the dielectric constant of a material, the larger the value of capacitor for the same area and spacing between the plates.

14 UNDERSTANDING CAPACITANCE

THE EFFECT OF CAPACITANCE ON AC VOLTAGE AND CURRENT

Voltage lags the current by 90° in a capacitor circuit.

You have already seen that sine waves of current and voltage are in phase through a resistance. You have also learned that current and voltage for an inductance are 90° out of phase, with the current lagging the voltage. Current and voltage are also 90° out of phase for a capacitance. As shown in *Figure 14-7*, the voltage lags the current by 90°.

Figure 14-7.
Voltage Lags Current for Capacitor

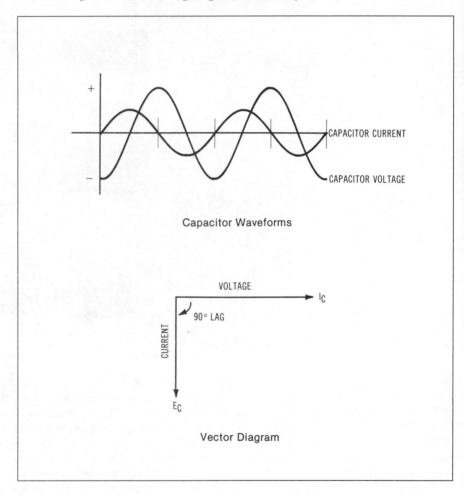

Capacitor Waveforms

Vector Diagram

224 UNDERSTANDING ELECTRICITY AND ELECTRONICS CIRCUITS

UNDERSTANDING CAPACITANCE

Used in ac circuits, capacitors return as much energy to the circuit as they absorb. The overall power dissipation of a capacitor is thus zero.

Capacitors offer opposition to current flow in ac circuits. This opposition is called capacitive reactance, and it is measured in units of ohms.

Capacitive reactance is inversely proportional to the frequency and value of capacitance.

Just like inductors, capacitors consume no power. During a sine-wave cycle, a capacitor absorbs energy during one quarter cycle, then returns it to the circuit during the next quarter cycle.

CAPACITIVE REACTANCE

Capacitors, like inductors, offer resistance to current flow in an ac circuit. This capacitive opposition to current flow in an ac circuit is called "capacitive reactance." Capacitive reactance is measured in ohms, and its mathematical symbol is X_C.

Just as inductive reactance depends on frequency and the value of inductance, capacitive reactance depends on frequency and capacitance. However the relationship between capacitive reactance, frequency, and capacitance is an inverse one. That is, the value of capacitive reactance decreases as frequency or capacitance increases. The equation for capacitive reactance looks like this:

$$X_C = \frac{1}{(2\pi fC)}$$

where,

X_C is the capacitive reactance in ohms,
π is the constant pi, 3.14,
f is the applied frequency in hertz,
C is the capacitance in farads.

The fact that values of capacitance are most often specified in units of microfarads means that you have to be careful about mixing units of measurement in the equation for capacitive reactance. If the frequency is in hertz and the capacitance is in microfarads, for example, the resulting capacitive reactance will be in units of megohms. *Table 14-1* summarizes the units of measurement for the most common combinations of units for frequency and capacitance.

**Table 14-1.
Units of Measurement
for Capacitive
Reactance Equation**

Frequency (f)	Capacitance (C)	Reactance X_C
hertz (Hz)	microfarads (μF)	megohms (MΩ)
kilohertz (kHz)	microfarads (μF)	kilohms (kΩ)
megahertz (MHz)	microfarads (μF)	ohms (Ω)
megahertz (MHz)	picofarads (pF)	megohms (MΩ)

To see how the equation and table work together, suppose you want to find the capacitive reactance of a circuit that has a capacitance of 0.1 µF and is operating at 4 kHz. According to the equation for capacitive reactance:

$$X_C = 1/(2\pi fC)$$
$$X_C = 1/(6.28 \times 4 \text{ kHz} \times 0.001 \text{ µF})$$
$$X_C = 39.8 \text{ Ω}$$

The frequency was specified in kilohertz and the capacitance in microfarads, so the result is in kilohms.

Current never flows between the plates of a capacitor, but charge and discharge current does flow through the connections leading to the plates of the capacitor. It is thus proper to speak in terms of the amount of current for a capacitor circuit.

If you have already calculated the capacitive reactance of such as circuit, you can apply Ohm's law to determine the amount of current: $I = E/X_C$. You will learn more about capacitive reactance, current, and voltage in the next chapters.

THE EFFECT OF CAPACITANCE ON PULSE WAVEFORMS

A capacitor tends to round off the edges of pulsed voltage waveforms and peak the edges of the corresponding current waveform.

When a sharp pulse, such as a square wave, is applied to a capacitive circuit, the capacitance opposes the sudden change in voltage. This results in a rounding off of the sudden voltage rise. Similarly, when the pulse voltage is suddenly removed, the voltage across the capacitor does not decrease suddenly, but trails off. Current is greatest when the change of voltage is greatest, so the current waveform will have a peak when the voltage rises suddenly, and another peak (but in the opposite direction) when the voltage drops, as shown in *Figure 14-8*.

There is always some resistance in a circuit. By choosing the right value of capacitance and resistance, you can assemble a circuit where the voltage takes a predetermined length of time to reach a certain value. This type of circuit provides a time-delay action that can be important for many kinds of circuits.

STRAY CAPACITANCE

Stray capacitance is unwanted capacitance between two conductors in a circuit.

Capacitive reactance decreases as frequency decreases. In equipment that uses very high frequencies (radio, television, and radar, for example), stray capacitance can present quite a problem.

In a transistor, antenna, or printed circuit board, there are always very small amounts of unwanted capacitance between adjacent conductors and between other objects that are supposed to be isolated from one another. "Stray capacitance" is any unwanted capacitance between two conductors.

**Figure 14-8.
Pulse Waveforms in
Capacitor Circuit**

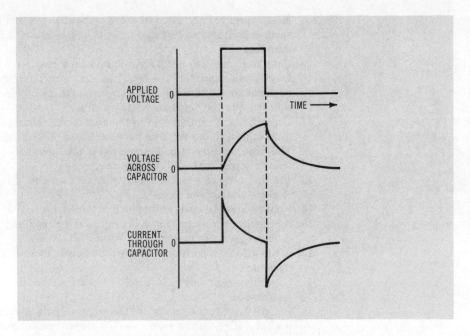

Stray capacitance is not usually important at the lower operating frequencies. But as the frequency increases, stray capacitive reactance decreases. This decrease in reactance can create unwanted paths for the signals, thereby reducing the level of the signals and possibly injecting them into other parts of the circuit where they are not supposed to appear.

The careful placement of components in very high-frequency circuits can greatly reduce the unwanted effects of stray capacitance.

WHAT HAVE WE LEARNED?

1. A basic capacitor consists of two metal plates separated by a dielectric material.
2. Current does not flow between the plates of a capacitor.
3. Current flows through a capacitor circuit (but not through the capacitor, itself) until the voltage on the plates of the capacitor is equal to that of the voltage source.
4. A capacitor can retain its charge after it is removed from its voltage source.
5. During the time a capacitor is being discharged, it acts as a source of voltage and current. A capacitor is completely discharged when there is no longer a difference in potential between its plates.
6. A capacitor in an ac circuit allows charge and discharge current to flow through the circuit.
7. Capacitors are rated in units of farads. One farad is the amount of capacitance required for allowing 1 ampere of current to change the charge at the rate of 1 volt per second.

8. Most capacitors are rated in microfarads (1 µF = 1-millionth of a farad). Very small capacitors can be rated in units of picofarads (1 pF = 1-millionth of a microfarad).
9. Capacitors are also rated according to their maximum operating voltage.
10. Electrolytic capacitors are polarized and must be connected so that their + terminal goes to the positive voltage source and their − terminal is connected to the negative voltage source.
11. Capacitance is proportional to the area of the plates and inversely proportional to the distance between them.
12. The dielectric strength of a material is the measure of its ability to withstand electric stress without breaking down.
13. The dielectric constant of a material is a measure of how well it passes electrostatic lines of force.
14. Voltage lags the current by 90° in a purely capacitive circuit.
15. A capacitor returns as much energy to the circuit as it absorbs. A capacitor does not consume power.
16. A capacitor's opposition to ac current flow is measured in terms of capacitive reactance.
17. Capacitive reactance is inversely proportional to the frequency and value of capacitance.
18. A capacitor tends to round off the edges of pulsed voltage waveforms and peak the edges of the corresponding current waveform.
19. Stray capacitance is unwanted capacitance between two conductors in a circuit.

KEY WORDS

Capacitive reactance
Capacitor
Dielectric
Dielectric constant
Dielectric strength
Electrolytic capacitor

UNDERSTANDING CAPACITANCE

Quiz for Chapter 14

1. A basic capacitor is composed of:
   a. two conductors separated by a dielectric.
   b. two dielectrics separated by a conductor.
   c. two coils separated by a dielectric.
   d. two coils separated by a conductor.

2. A dielectric material is:
   a. a good conductor.
   b. a good insulator.
   c. a good conductor of magnetic fields.
   d. a poor conductor of electrostatic fields.

3. Which one of the following statements is true?
   a. Pushing electrons onto one plate of a capacitor attracts electrons onto the other plate.
   b. Pushing electrons onto one plate of a capacitor forces electrons off the other plate.

4. Which one of the following statements is true?
   a. Current flows in a capacitor circuit only while the charge on the plates is large enough to force the dielectric to conduct electrons.
   b. Current flows in a capacitor circuit only while ions are permitted to flow through the dielectric material.
   c. Current can flow in a capacitor circuit, but not through the dielectric material itself.
   d. Power returned to the circuit is always greater than the power absorbed.

5. A farad is defined as the amount of capacitance necessary for:
   a. causing an ac phase shift greater than 90°.
   b. dissipating 1 W of power.
   c. storing 1 V for 1 second.
   d. changing the voltage on the plates at the rate of 1 V per second when 1 A of current is flowing.
   e. changing the current at the rate of 1 A per second when 1 V is applied.

6. 1 microfarad (μF) is the same as:
   a. 1-millionth of a picofarad.
   b. 1000 millifarads.
   c. 1 million picofarads.
   d. 1 million farads.

7. The value of a capacitor can be made larger by:
   a. increasing the area of the plates.
   b. decreasing the area of the plates.
   c. increasing the voltage applied to the plates.
   d. increasing the frequency of the applied voltage.
   e. decreasing the frequency of the applied voltage.

8. The value of a capacitor can be made larger by:
   a. moving the plates further apart.
   b. moving the plates closer together.
   c. decreasing the voltage applied to the plates.
   d. increasing the voltage applied to the plates.
   e. decreasing the frequency of the applied voltage.

9. The dielectric strength of a material is a measure of:
   a. the amount of voltage it can withstand before breaking down.
   b. how well the material concentrates electrostatic lines of force.
   c. how well the material concentrates magnetic lines of force.
   d. the amount of opposition to current flow.

10. The dielectric constant of a material is a measure of:
    a. the amount of voltage it can withstand before breaking down.
    b. how well the material concentrates electrostatic lines of force.
    c. how well the material concentrates magnetic lines of force.
    d. the amount of opposition to current flow.

11. In a purely capacitive ac circuit:
    a. voltage leads the current by 90°.
    b. voltage lags the current by 90°.

12. Increasing the frequency of the ac waveform applied to a capacitor:
    a. increases its capacitive reactance.
    b. decreases its capacitive reactance.
    c. has no effect on its capacitive reactance.

13. Increasing the value of capacitance in a circuit:
    a. increases its value of capacitive reactance.
    b. decreases its value of capacitive reactance.
    c. has no effect on its capacitive reactance.

14. What is the capacitive reactance of an 0.01-μF capacitor that is operating in a 1000-Hz circuit?
    a. 62.8 Ω.
    b. 15.9 kΩ.
    c. 62.8 kΩ.
    d. 1.59 MΩ.

15. When applying a square waveform to a capacitor, the circuit tends to:
    a. peak the voltage and round off the current waveforms.
    b. peak the current and round off the voltage waveforms.
    c. peak both the voltage and current waveforms.
    d. round off both the voltage and current waveforms.

15

Capacitive Circuits

ABOUT THIS CHAPTER

Chapter 14 described the basic elements of capacitance. We saw that a capacitor attempts to oppose any change in voltage across it and that this opposition can be measured in terms of capacitive reactance. In this chapter, you will learn more about capacitors and capacitive circuits, including how to calculate the total capacitance and total capacitive reactance of various combinations of capacitors—series, parallel, and combinations of series and parallel circuits. You will also see how it is possible to determine the current through capacitive circuits and the voltage drop across them.

A PURELY CAPACITIVE CIRCUIT

A purely capacitive circuit contains only capacitances and capacitive reactance provides the only opposition to current flow. We have already seen how an ac voltage applied to a capacitor lags the current—the charge and discharge current—by 90°. That is true as long as the circuit is purely capacitive. While there is actually no such thing as a purely capacitive circuit, the concept of a pure capacitor is an important one that you will use for analyzing the operation of practical circuits that combine capacitance, resistance, and inductance.

So the voltage applied to a purely capacitive circuit lags the current by exactly 90°. There is a different way to state the same idea: the current through a purely capacitive circuit leads the voltage by exactly 90°. *Figure 15-1* contrasts these two statements by vector diagram. Both ideas are correct, but you are going to find that it is often more convenient to think of capacitive circuits in terms of leading current rather than lagging voltage.

In a purely capacitive circuit, the only opposition to current flow is the capacitive reactance of the capacitors.

In a purely capacitive circuit, current leads the voltage by exactly 90°.

15 CAPACITIVE CIRCUITS

Figure 15-1.
Purely Inductive Circuit: Current and Voltage Separated by 90°.

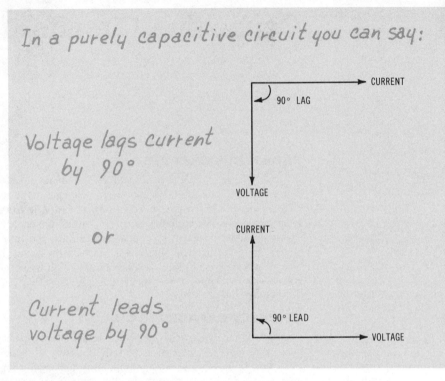

You have previously learned that the capacitive reactance of a capacitor is determined by the value of capacitance and the operating frequency of the ac source. The mathematical expression for this relationship is:

$$X_C = \frac{1}{(2\pi fC)}$$

where,

X_C is the capacitive reactance in ohms,
f is the operating frequency in hertz,
C is the value of the capacitor in farads,
π is the constant 3.14.

Like resistors, two or more capacitors can be connected with one another in a variety of ways. Capacitors can be connected in series, parallel, and combinations of series and parallel.

OHM'S LAW APPLIED TO CAPACITIVE CIRCUITS

The basic form of Ohm's law states that the amount of current flowing through a circuit is equal to the applied voltage divided by the opposition to current flow. According to this important principle, larger

CAPACITIVE CIRCUITS

15

values of voltage cause larger amounts of current to flow. On the other hand, increasing the amount of opposition to current flow decreases the amount of current.

Ohm's law applies to purely capacitive circuits: $I = E/X_C$

When you are working with circuits that contain nothing but pure resistances, the resistors provide the opposition to current flowing through the circuits. When working with purely inductive circuits, inductive reactances provide the opposition to current flow. And when working with purely capacitive circuits, the capacitive reactance of the capacitors provides the opposition to current flow. Ohm's law applies to purely capacitive circuits, where the resistance term (R) is replaced with capacitive reactance (X_C).

Ohm's law for purely capacitive circuits thus looks like this:

$$I = \frac{E}{X_C}$$

where,

I is the amount of current flowing through the capacitance in amperes (usually rms),
E is the voltage applied to the capacitor in volts (usually rms),
X_C is the capacitive reactance of the capacitor in ohms.

Figure 15-2 shows a circuit composed of an ac power source and a single capacitor. The source is rated at 12 V ac, 100 Hz. The value of the capacitor is 0.1 µF. What is the current through this circuit? The capacitive-reactance version of Ohm's law provides a means for calculating the current, but first, as shown in Step 1, you must determine the value of capacitive reactance. The units are given in hertz and microfarads, so the immediate result is in units of megohms. In this case, the result can be better expressed in units of kilohms.

**Figure 15-2.
Determining Current
through Simple
Capacitive Circuit**

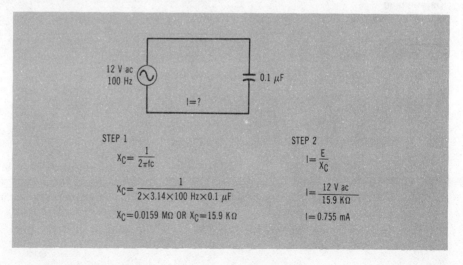

UNDERSTANDING ELECTRICITY AND ELECTRONICS CIRCUITS

15 CAPACITIVE CIRCUITS

Now that you know that the amount of capacitive reactance is 15.9 kilohms, you can apply the capacitive-reactance version of Ohm's law to determine the current through the capacitor. This calculation, shown in Step 2, is in units of volts and kilohms, so the result is in milliamperes.

Table 15-1 summarizes three forms of the basic capacitance equation and Ohm's law as it applies to ideal capacitive circuits. Consider the information supplied for the capacitor circuit in *Figure 15-3*. You know the applied ac voltage, the operating frequency, and the desired amount of ac current. The problem is to determine the value of capacitance required to do the job.

**Table 15-1.
Summary of Equations for Capacitive Reactance**

Inductance Equations	Ohm's Law Equations (pure capacitance)
$X_C = 2\pi f C$	$I = E/X_C$
$C = X_C/2\pi f$	$E = IX_C$
$f = X_C/2\pi C$	$X_C = E/I$

X_C = capacitive reactance in ohms
C = capacitance in farads
f = frequency in hertz
I = current in amperes (usually rms)
E = voltage in volts (usually rms)

**Figure 15-3.
Determining Capacitance**

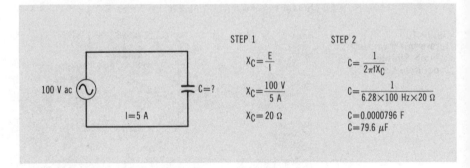

STEP 1
$X_C = \dfrac{E}{I}$
$X_C = \dfrac{100\ V}{5\ A}$
$X_C = 20\ \Omega$

STEP 2
$C = \dfrac{1}{2\pi f X_C}$
$C = \dfrac{1}{6.28 \times 100\ Hz \times 20\ \Omega}$
$C = 0.0000796\ F$
$C = 79.6\ \mu F$

Table 15-1 shows that you can calculate the value of capacitance in such a circuit by $C = 1/(X_C/2\pi f)$. However, you do not yet know the value of capacitive reactance, X_C. So the procedure begins by using the form of Ohm's law that provides the amount of capacitive reactance in terms of voltage and current:

CAPACITIVE CIRCUITS

15

$$X_C = \frac{E}{I}$$

$$X_C = \frac{100}{5}$$

$$X_C = 20 \; \Omega$$

Knowing the value of X_C, you can now apply the equation that solves for the value of capacitance as shown in Step 2. Because the units are in hertz and ohms, the result is in farads. The result is a very small decimal value when expressed in units of farads, so you should multiply by 1 million to convert it to microfarads. The conclusion is that a 79.6 μF capacitance will do the job for you.

DC POWER EQUATIONS APPLIED TO CAPACITIVE CIRCUITS

You have already learned that the electrical energy supplied to a purely resistive circuit is dissipated in the form of heat. The wattage rating of a resistor must be selected so that it can handle the amount of power that is to be dissipated in that fashion. But you have also seen that a pure inductor does not dissipate heat. Energy absorbed by an inductor is used to build magnetic lines of force, and the same energy ultimately is returned to the circuit. There is no overall power loss in a purely inductive circuit.

There is no power loss in a purely capacitive circuit.

A purely capacitive circuit does not dissipate heat, either. When a capacitor is being charged, the electrical energy is stored in the form of electrical charges on the plates of the capacitor. And when a capacitor is discharging, it returns that stored energy to the circuit. There is no overall power loss in a purely capacitive circuit.

Figure 15-4 shows the sine waveforms for current, voltage, and power in purely resistive, inductive, and capacitive circuits. There are no phase differences for the resistive waveforms. The only differences between inductive and capacitive waveforms are the directions of phase shift.

During the time the capacitive power waveform is positive, the capacitor absorbs energy from the circuit in order to charge its plates. During the time the power waveform is negative, the capacitor returns its charge to the circuit. An ideal capacitor always returns as much energy as it absorbs. The overall result is that it consumes no power at all. The power equation thus has no real meaning when applied to a purely capacitive circuit.

UNDERSTANDING ELECTRICITY AND ELECTRONICS CIRCUITS

**Figure 15-4.
Voltage, Current, and
Power Waveforms**

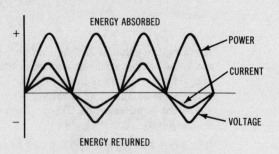

a. Purely Resistive Circuit

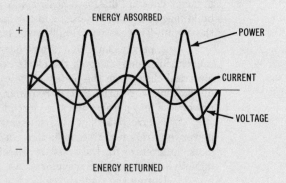

b. Purely Capacitive Circuit

ANALYZING CAPACITORS CONNECTED IN PARALLEL

The leads for any number of capacitors can be connected to make up a parallel capacitor circuit. *Figure 15-5* shows three capacitors connected in parallel. The circuit is purely capacitive because it shows no other kind of component that can provide opposition to current flow. The only opposition to current flow in this circuit comes from the combined capacitive reactance of the capacitors.

CAPACITIVE CIRCUITS

**Figure 15-5.
Parallel Capacitor
Circuit**

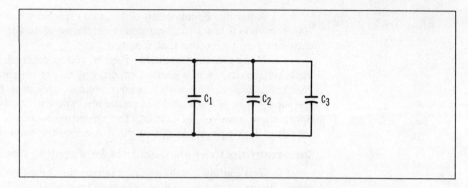

The same voltage is applied to each component in a parallel capacitor circuit. If the circuit is purely capacitive, the total current is divided among the capacitors according to their values of reactance.
The total capacitance of two or more capacitors connected in parallel is equal to the sum of the individual capacitor values.

Like any other kind of parallel circuit, the same voltage appears across each capacitor in a parallel capacitor circuit. And if the circuit is purely capacitive, the total current divides among the capacitors according to their values of reactance.

Determining the Total Capacitance

The total capacitance of two or more capacitors connected in parallel is equal to the sum of their individual capacitances. If the total capacitance is shown as C_T and the individual capacitance values as C_1, C_2, C_3, and so on, the equation for determining the total capacitance of a parallel capacitor circuit is:

$$C_T = C_1 + C_2 + C_3...$$

Thus, a parallel circuit composed of three capacitors rated at 1 μF, 4 μF, and 3 μF thus has a total capacitance of $1 + 4 + 3$, or 8 μF.

Determining Capacitive Reactances in a Parallel Circuit

Ac current flows through each branch of a parallel capacitive circuit. The amount of current in each branch depends on the capacitive reactance of that branch—the larger the amount of capacitive reactance, the smaller the amount of current.

Use the product-over-sum rule to determine the total capacitive reactance of two capacitors connected in parallel.

The equations for determining the total reactance of a parallel circuit have the same general form as those for determining the total resistance and total inductive reactance in parallel circuits. The general equation is:

$$X_{CT} = \frac{1}{[(1/X_{C1}) + (1/X_{C2}) + (1/X_{C3})...]}$$

and the product-over-sum version is:

$$X_{CT} = \frac{(X_{C1} \times X_{C2})}{(X_{C1} + X_{C2})}$$

15 CAPACITIVE CIRCUITS

A parallel circuit composed of two capacitors having reactances of 100 Ω and 800 Ω has a total capacitive reactance of 44.4 Ω. Use the product-over-sum rule to confirm that result.

Many practical situations require you to determine the total capacitive reactance of a parallel circuit, but do not provide the amounts of reactance. You might know the source voltage, operating frequency, and the capacitance of the individual capacitors. You can use the available information, however, to calculate the capacitive reactance of each capacitor. Having done that, you can determine the total reactance.

Determining Branch Currents in Parallel Capacitive Circuits

You can always determine the current through a capacitor in an ac circuit by applying a capacitive-reactance form of Ohm's law, $I = E/X_C$. That formula can be applied to any capacitor in a parallel circuit, assuming you know the amount of voltage dropped across it. And in simple parallel circuits, the voltage across each capacitor is the same as the source voltage.

So if capacitor C_1 in a parallel circuit has a capacitive reactance of 250 Ω and the source voltage is 100 V ac, the current through the capacitor is:

$$I_{C1} = \frac{E}{X_{C1}}$$

$$I_{C1} = \frac{100}{250}$$

$$I_{C1} = 0.4 \text{ A}$$

In most practical situations, however, you are provided with the source voltage, the operating frequency, and the capacitance values of the individual capacitors. None of this information fits directly into the capacitive-reactance version of Ohm's law for current. You should use the available information to determine the capacitive reactance of each capacitor and then apply Ohm's law to find the current through each capacitor. *Figure 15-6* illustrates the procedure for a specific circuit.

CAPACITIVE CIRCUITS

**Figure 15-6.
Analysis of Parallel
Capacitor Circuit**

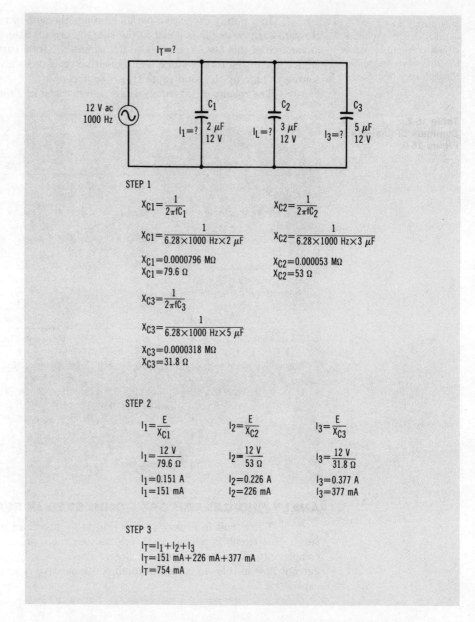

In Step 1, you determine the capacitive reactance of each capacitor in the circuit. Knowing the reactance of each capacitor lets you apply Ohm's law to determine the current through each one of them. This is calculated in Step 2.

15 CAPACITIVE CIRCUITS

In a purely capacitive parallel circuit, the total circuit current is equal to the sum of the currents through each branch.

In a purely capacitive parallel circuit, the sum of the currents through each capacitor is equal to the source current. Step 3 takes advantage of this fact. You can also determine the total current by applying Ohm's law to the total voltage and capacitive reactance—dividing the source voltage by the total capacitive reactance.

The results of this analysis are summarized in *Table 15-2*.

Table 15-2. Summary of Analysis of Figure 15-6

Total Values		Values for C_1	
Total circuit voltage: E_T = 12.6 V	(Given)	Capacitance of C_1: C_1 = 1 μF	(Given)
Operating frequency: f = 400 Hz	(Given)	Capacitive reactance of C_1: X_{C1} = 2512 Ω	(Step 3)
Total capacitance: C_T = 1.33 μF	(Step 2)	Current through C_1: I_{C1} = 0.002 A	(Step 4)
Total capacitive reactance X_{CT} = 5850 Ω	(Step 3)	Voltage across C_1: E_{C1} = 5.02 V	(Step 4)
Total circuit current I_T = 0.0022 A	(Step 3)		
Values for C_2		**Values for C_3**	
Capacitance of C_2: C_2 = 2 μF	(Given)	Capacitance of C_3: C_3 = 4 μF	(Given)
Capacitive reactance of C_2: X_{C2} = 5024 Ω	(Step 5)	Capacitive reactance of C_3: X_{C3} = 10,048 Ω	(Step 5)
Current through C_2: I_{C2} = 0.0015 A	(Step 5)	Current through C_3: I_{C3} = 0.0015 A	(Step 5)
Voltage across C_2: E_{C2} = 7.58 V	(Step 4)	Voltage across C_3: E_{C3} = 7.58 V	(Step 4)

ANALYZING CAPACITORS CONNECTED IN SERIES

The circuit in *Figure 15-7* shows three capacitors connected in series. The circuit is purely capacitive because there are no other kinds of components offering opposition to current flow. The only opposition to current flow in this circuit is the combined capacitive reactance of the capacitors.

CAPACITIVE CIRCUITS

**Figure 15-7.
Series Capacitor Circuit**

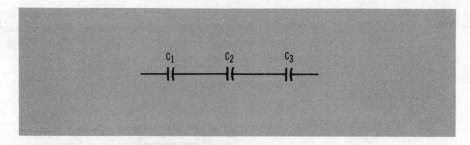

Like any other kind of series circuit, the same current flows everywhere in the circuit. And if the circuit is purely capacitive, the total voltage is divided among the capacitors according to their values of reactance. So, capacitors having the larger reactance values will have larger voltage drops.

Determining Total Capacitance in a Series Circuit

The total capacitance of two or more capacitors connected in series is less than the value of the smallest capacitor. The general equation for finding total series capacitance is:

$$C_T = \frac{1}{[(1/C_1) + (1/C_2) + (1/C_3)...]}$$

An equation of that complexity might serve a useful purpose under some circumstances, but most technicians prefer the product-over-sum version. Given two capacitors connected in series, their total capacitance can be determined by:

$$C_T = \frac{R_1 R_2}{(R_1 + R_2)}$$

where,

C_T is the total capacitance of the circuit,
C_1 is the capacitance of one of the capacitors,
C_2 is the capacitance of the second capacitor.

Suppose you have a parallel circuit composed of two capacitors rated at 1 μF and 4 μF. You can find the total capacitance this way:

$$C_T = R_1 R_2/(R_1 + R_2)$$
$$C_T = 1\ \mu F \times 4\ \mu F/(1\ \mu F + 4\ \mu F)$$
$$C_T = 4\ \mu F/5\ \mu F$$
$$C_T = 0.8\ \mu F$$

The same current flows everywhere in a series capacitor circuit. If the circuit is purely capacitive, the total voltage is divided among the capacitors according to their values of reactance.

The total capacitance of two or more capacitors connected in series is less than the value of the smallest capacitor.

15   CAPACITIVE CIRCUITS

Figure 15-8 illustrates the procedure for using the product-over-sum rule to determine the total capacitance of a circuit composed of more than two capacitors connected in series. The key to the procedure is to work with the values, two at a time, to create equivalent capacitances.

**Figure 15-8.
Total Capacitance of Series Circuit**

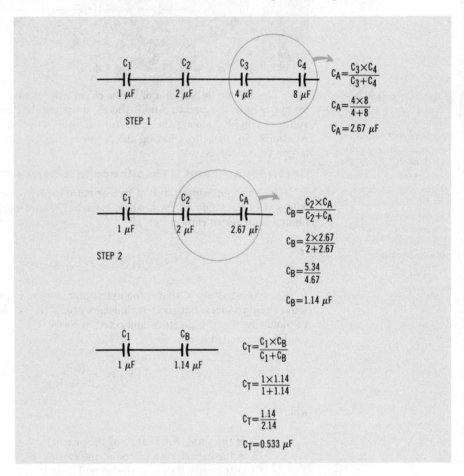

Determining Capacitive Reactances in a Series Circuit

Ac voltage is dropped across each capacitor in a series circuit. The amount of voltage across each capacitor depends on its own capacitive reactance and the total capacitive reactance of the circuit.

Recall that the total resistance of a series resistor circuit is equal to the sum of the individual resistances. You'll also recall that the total reactance of a purely inductive series circuit is equal to the sum of the individual inductive reactances. The same general principle applies to series capacitor circuits: the total capacitive reactance of a series capacitor circuit is equal to the sum of the individual reactances. The mathematical expression is:

The total capacitive reactance of a series capacitor circuit is equal to the sum of the individual reactances.

CAPACITIVE CIRCUITS

15

$$X_{CT} = X_{C1} + X_{C2} + X_{C3}...$$

If you have a series capacitor circuit in which the individual reactances are 100 Ω, 50 Ω, and 20 Ω, the total inductive reactance is 100 + 50 + 20 = 170 Ω.

Many practical situations require that you determine the total capacitive reactance of a parallel circuit, but do not provide the amounts of reactance. You might know the source voltage, the operating frequency and the capacitance of the individual capacitors. If so, you can use that information to calculate the capacitive reactance of each capacitor. And having done that, you can determine the total reactance. *Figure 15-9* illustrates this procedure.

Figure 15-9.
Total Reactance of Series Capacitor Circuit

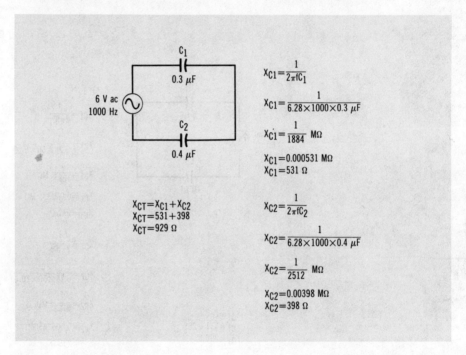

Determining Voltage Drops in Series Capacitive Circuits

You can determine the ac voltage drop across a capacitor in a series circuit by applying this form of Ohm's law:

$$E_C = IX_C$$

where,

E_C is the voltage drop across the capacitor in volts,
I is the total series-circuit current in amperes,
X_C is the capacitive reactance of the capacitor in ohms.

UNDERSTANDING ELECTRICITY AND ELECTRONICS CIRCUITS 243

15 CAPACITIVE CIRCUITS

Using that equation requires values for capacitive reactance and current, and you often have to calculate these values before you can apply the equation.

Figure 15-10 shows a complete analysis of a series capacitor circuit based on applied voltage, applied frequency, and the values of the capacitors.

**Figure 15-10.
Analysis of Series
Capacitor Circuit**

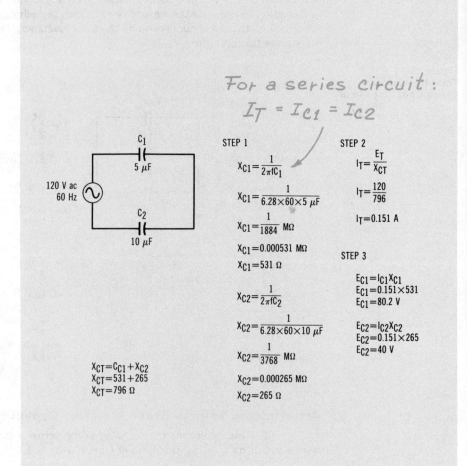

CAPACITIVE CIRCUITS

15

In Step 1, you determine the values of capacitive reactance. In Step 2, you use the results to determine the amount of ac current flowing through the circuit. Step 3 uses this value of current and the values of inductive reactance to calculate the voltage drops across the individual capacitors. The results of the analysis are shown in *Table 15-3*.

Table 15-3. Summary of Analysis of Figure 15-10

Total Values		Values for C_1	
Total circuit voltage: $E_T =$ 120 V	(Given)	Capacitance of C_1: $C_1 =$ 5 µF	(Given)
Operating frequency: $f =$ 60 Hz	(Given)	Capacitive reactance of C_1: $X_{C1} = 531\ \Omega$	(Step 1)
Total capacitive reactance $X_{CT} = 796\ \Omega$	(Step 1)	Current through C_1: $I_{C1} =$ 0.151 A	(Step 2)
Total circuit current $I_T =$ 0.151 A	(Step 2)	Voltage across C_1: $E_{C1} =$ 80.2 V	(Step 3)
Values for C_2			
Capacitance of C_2: $C_2 =$ 10 µF	(Given)		
Capacitive reactance of C_2: $X_{C2} = 265\ \Omega$	(Step 1)		
Current through C_2: $I_{C2} =$ 0.151 A	(Step 2)		
Voltage across C_2: $E_{C2} =$ 40 V	(Step 3)		

WHAT HAVE WE LEARNED?

1. In a purely capacitive circuit, the only opposition to current flow is the capacitive reactance of the capacitors.
2. The current in a purely capacitive circuit leads the voltage by exactly 90°.
3. There is no such thing as a pure capacitor nor a purely capacitive circuit; however, the principles are important to the analysis of ac circuits that include capacitors.
4. Ohm's law applies to purely capacitive circuits. The opposition to current flow in this case is capacitive reactance rather than resistance or inductive reactance.
5. Dc power equations have no practical application in purely capacitive circuits.
6. The total capacitance of capacitors connected in parallel is equal to the sum of the individual capacitance values.
7. The same voltage is found across all branches in a parallel capacitor circuit.
8. In a purely capacitive parallel circuit, the sum of the currents for the capacitors is equal to the source current.

9. You can use the product-over-sum rule to determine the total capacitive reactance of a parallel circuit.
10. Capacitors connected in series have the same current.
11. You can use the product-over-sum rule to determine the total capacitance of two capacitors connected in series. The total capacitance of such a circuit is less than the value of the smaller capacitance.
12. In a purely capacitive series circuit, the total voltage is equal to the sum of the voltages across the individual branches.
13. The total capacitive reactance of a series capacitor circuit is equal to the sum of the individual reactances.

CAPACITIVE CIRCUITS

15

Quiz for Chapter 15

1. Which one of the following phrases most accurately describes a purely capacitive circuit?
   a. Resistances provide the only opposition to current flow.
   b. Capacitive reactance provides the only opposition to current flow.
   c. Combinations of resistance and capacitive reactance provide any opposition to current flow.
   d. The ac voltage leads the current by 90°.
   e. The ac current and voltage are in phase.

2. For a pure capacitor:
   a. ac current and voltage are exactly in phase.
   b. ac current leads the voltage by 90°.
   c. ac current lags the voltage by 90°.
   d. ac current is converted to dc voltage.

3. Which one of the following statements is true?
   a. Capacitive reactance is at a minimum for dc source voltages.
   b. Capacitive reactance decreases as the operating frequency increases.
   c. Capacitive reactance increases as the operating frequency increases.
   d. There is no meaningful relationship between capacitive reactance and operating frequency.

4. What is the capacitive reactance of a 0.1-μF capacitor that is operating at 1000 Hz?
   a. Less than 1 Ω.
   b. 100 Ω.
   c. 314 Ω.
   d. 628 Ω.
   e. 1000 Ω.
   f. 1590 Ω.

5. How much current flows through a 0.02-μF capacitor that is operating from a 12-V ac, 100-Hz source?
   a. 0.151 mA.
   b. 157 mA.
   c. 0.995 mA.
   d. 2.02 A.
   e. 10 A.
   f. 100 A.

6. Doubling the operating frequency of a capacitive circuit:
   a. has no effect on the capacitive reactance.
   b. doubles the amount of capacitive reactance.
   c. cuts the capacitive reactance in half.
   d. multiplies the capacitive reactance by 6.28.

7. Doubling the operating frequency of a purely capacitive circuit:
   a. has no effect on the total circuit current.
   b. doubles the amount of total current.
   c. cuts the total current by one-half.

UNDERSTANDING ELECTRICITY AND ELECTRONICS CIRCUITS

15 CAPACITIVE CIRCUITS

8. Doubling the operating voltage of a purely capacitive circuit:
   a. has no effect on the capacitive reactance.
   b. doubles the amount of capacitive reactance.
   c. cuts the capacitive reactance in half.
   d. multiplies the capacitive reactance by 6.28.

9. Which one of the following statements is true?
   a. Power dissipation of a pure capacitor decreases with operating frequency.
   b. Power dissipation of a pure capacitor increases with operating frequency.
   c. There is no meaningful relationship between the power dissipation of a pure capacitor and its operating frequency.

10. The total capacitance of a series capacitor circuit is:
    a. equal to the sum of the individual capacitance values.
    b. equal to the sum of the individual capacitive-reactance values.
    c. equal to the source voltage divided by total current.
    d. less than the value of the smallest capacitor.

11. The total capacitive reactance of a parallel capacitor circuit is:
    a. equal to the sum of the individual capacitance values.
    b. equal to the sum of the individual capacitive-reactance values.
    c. equal to the source voltage divided by total current.
    d. less than the capacitance value of the smallest capacitor.

Figure 15-11.
Circuit for Questions 12–16

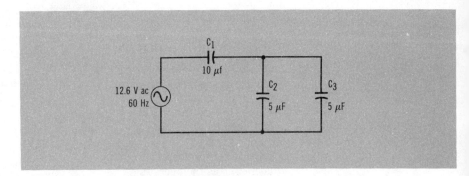

12. What is the total capacitance of the circuit in *Figure 15-11*?
    a. 5 µF.
    b. 10 µF.
    c. 12.5 µF.
    d. 15 µF.
    e. 20 µF.

13. What is the total current for the circuit in *Figure 15-11*?
    a. 23.7 mA.
    b. 47.5 mA.
    c. 59.4 mA.
    d. 71.3 mA.
    e. 95 mA.

CAPACITIVE CIRCUITS

15

14. What is the voltage drop across capacitor C_1 in *Figure 15-11*?
    a. 1.26 V.
    b. 4.2 V.
    c. 6.3 V.
    d. 10.5 V.
    e. 12.6 V.

15. What is the voltage drop across capacitor C_2 in *Figure 15-11*?
    a. 1.26 V.
    b. 4.2 V.
    c. 6.3 V.
    d. 10.5 V.
    e. 12.6 V.

16. What is the voltage drop across capacitor C_3 in *Figure 15-11*?
    a. 1.26 V.
    b. 4.2 V.
    c. 6.3 V.
    d. 10.5 V.
    e. 12.6 V.

Understanding RC Circuits

ABOUT THIS CHAPTER

In Chapter 15, we analyzed purely capacitive circuits. This chapter shows you how to deal with circuits that are not purely capacitive. You will learn about the distribution of currents and voltages in circuits that contain both resistances and capacitances. You will calculate phase angles and impedances.

Many of the principles used in this chapter have already been described in connection with RL circuits in Chapter 13. If that material is a little hazy, you may wish to review it before attempting to understand how vector diagrams and certain mathematical principles apply to RC circuits.

RC CIRCUITS

An RC circuit is one that contains both resistance and capacitance.

Chapters 14 and 15 described capacitance and capacitive circuits only in terms of pure capacitances. In these circuits, capacitive reactance provides the only opposition to current flow and ac voltage lags the applied current by exactly 90°. Whenever you include a resistance in a capacitor circuit, the circuit is no longer purely capacitive—resistance and capacitive reactance both provide opposition to current flow, and the ac voltage for the circuit lags the total current by some amount less than 90°. A circuit that includes both resistance (R) and capacitance (C) is called an RC circuit. *Figure 16-1* illustrates some examples of RC circuits.

You will still use what you know about pure capacitors to analyze RC circuits, but RC circuits require a more extensive analysis.

WORKING WITH SERIES RC CIRCUITS

One of the important characteristics shared by any kind of series circuit is that the same current flows through every part of the circuit. On the other hand, the sum of the voltage drops across the components in a series RC circuit does not equal the source voltage as in purely capacitance circuits.

16

UNDERSTANDING RC CIRCUITS

Figure 16-1.
RC Circuits

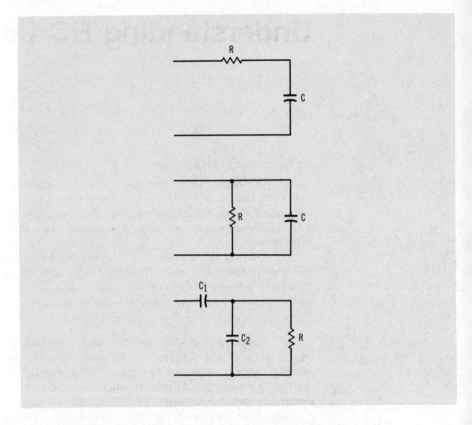

The same current flows through every component in a series circuit. This means the amplitude through every component is equal to the amplitude of the total current:

$$I_T = I_C = I_R$$

where,

I_T is the total current in amperes,
I_C is the current through the capacitance in amperes,
I_R is the current through the resistances in amperes.

The fact that the same current flows through every component in a series circuit also means that the phase of the current is the same through each component.

Figure 16-2 illustrates these important ideas about current through the components in a series RC circuit. The sine-wave diagrams show that the currents through the resistance and capacitance have the same amplitude and phase as the total current. The current vector diagram shows the three current vectors superimposed. They are identical.

**Figure 16-2.
Currents in Series RC Circuits**

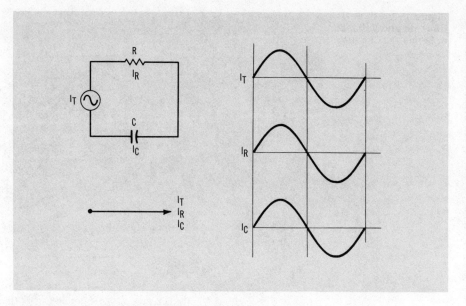

Voltage in a Series RC Circuit

The amplitude of the voltage drop across a resistance is equal to the amplitude of the current multiplied by the value of the resistance: $E_R = I_R \times R$. Furthermore, the voltage and current waveforms are exactly in phase, as shown by the resistor waveforms in *Figure 16-3b*.

The voltage drop across a capacitance is equal to the amplitude of the current through the capacitance multiplied by the value of capacitive reactance. The voltage across the capacitance, however, lags the current by exactly 90°.

The equation for the voltage drop across a capacitor is:

$$E_C = I_C \times X_C$$

where,

E_C is the amplitude of the voltage across the capacitance in volts,
I_C is the amplitude of the current through the capacitance in amperes,
X_C is the capacitive reactance of the capacitance in ohms.

Regarding the 90° phase difference between the current and voltage for a capacitance, you have already learned that the voltage lags the current by 90°. This is expressed by the capacitor waveform in *Figure 16-3c*.

16 UNDERSTANDING RC CIRCUITS

**Figure 16-3.
Currents and Voltages
in Series RC Circuits**

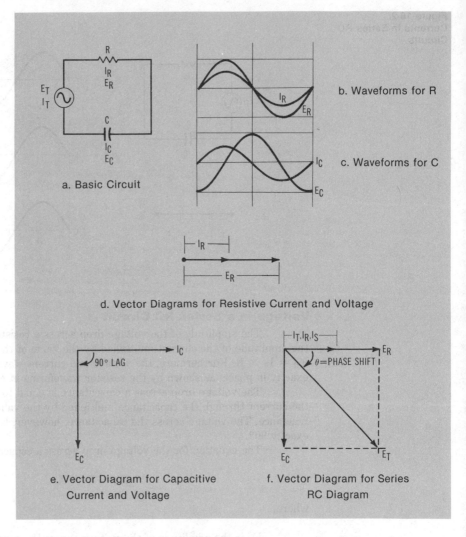

The voltage across the capacitor in a series RC circuit lags the voltage across the resistance by 90°.

Comparing the waveforms for the resistance and capacitance in a series RC circuit, you can see that the currents are in phase. This is true for any series circuit. The voltages, however, are 90° out of phase. The voltage across the capacitance lags the voltage across the resistance by 90°.

The separate vector diagrams for the resistance and capacitance also show the phase relationships for those components. The vector diagram for the resistance shows the voltage and current vectors pointing in the same direction, indicating that the current and voltage for the resistance are in phase. The vector diagram for the capacitance, however, shows the capacitive voltage lagging the capacitive current by 90°.

254 UNDERSTANDING ELECTRICITY AND ELECTRONICS CIRCUITS

UNDERSTANDING RC CIRCUITS 16

Aligning the current vectors for the resistor and capacitor produces an overall current/voltage vector diagram for series RC circuits. *Figure 16-3f* shows that the total voltage, E_T, is the diagonal of a rectangle having the resistive and capacitive voltages as its sides. The length of the E_T vector indicates the amplitude of the total voltage, and the phase angle indicates the amount and direction of phase shift.

In a series RC circuit, the total voltage lags the total current by less than 90°.

The series RC vector diagram in *Figure 16-4* includes all the information required for understanding and analyzing series RC circuits. Notice, for example, that the total voltage lags the total current by some angle less than 90°.

**Figure 16-4.
Current and Voltage Vector Diagram for Series RC Circuits**

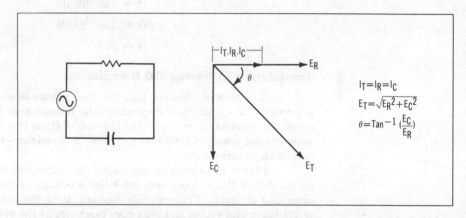

$I_T = I_R = I_C$

$E_T = \sqrt{E_R^2 + E_C^2}$

$\theta = \text{Tan}^{-1}\left(\dfrac{E_C}{E_R}\right)$

The total voltage in a series RC circuit is greater than either the resistive or capacitive voltage, but less than the sum of the two.

The diagram also shows that the amplitude of the total voltage will always be greater than either the resistive voltage or the capacitive voltage but less than the sum of the two.

You can closely estimate the total voltage and phase angle by applying the graphing procedures described for series RL circuits in Chapter 13. But if you are prepared to handle the mathematics, you can get more accurate results by applying some equations.

Use this equation to find the amplitude of the total voltage.

$$E_T = \sqrt{E_R^2 + E_C^2}$$

Suppose that the voltage across the resistance is 4 V and the voltage across the capacitance is 3 V. Using the equation for the total voltage in a series RC circuit:

$$E_T = \sqrt{E_R^2 + E_C^2}$$
$$E_T = \sqrt{4^2 + 3^2}$$
$$E_T = \sqrt{16 + 9}$$
$$E_T = \sqrt{25}$$
$$E_T = 5 \text{ V}$$

Calculating the value of the phase angle, Θ, is a matter of applying the following Arctangent equation:

$$\Theta = \text{Tan}^{-1}(E_C/E_R)$$

If $E_C = 3$ V and $E_R = 4$ V, as in the previous sample, the phase angle can be found this way:

$$\Theta = \text{Tan}^{-1}(E_C/E_R)$$
$$\Theta = \text{Tan}^{-1}(3/4)$$
$$\Theta = \text{Tan}^{-1}(0.75)$$
$$\Theta = 36.9°$$

Impedance in Series RC Circuits

You have just learned that the total voltage in a series RC circuit is greater than either the voltage across the resistance or the voltage across the capacitance but less than the sum of those two voltages. The same general principle applies to the circuit's impedance—its total opposition to current flow.

Figure 16-5 shows an impedance vector diagram for a series RC circuit. Notice that it looks very much like a voltage vector diagram for the same kind of circuit. However, this diagram shows the value of resistance as the horizontal vector and capacitive reactance as the vertical vector. The diagonal of the rectangle formed by the resistance and reactance vectors is the impedance vector, Z. The length of the impedance vector specifies the value of impedance, and the angle of the impedance vector with respect to the resistance vector represents the phase angle.

**Figure 16-5.
Impedance Vector Diagram for Series RC Circuits**

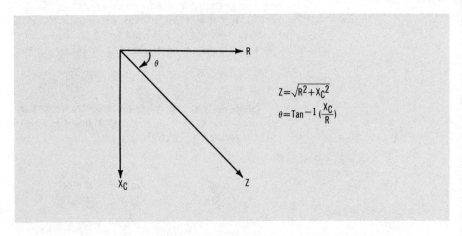

$$Z = \sqrt{R^2 + X_C^2}$$
$$\theta = \text{Tan}^{-1}\left(\frac{X_C}{R}\right)$$

You can estimate the amount of impedance and the phase angle for the circuit by graphical methods, or you can use the following equations:

UNDERSTANDING RC CIRCUITS

$$Z = \sqrt{R^2 + X_C^2}$$

where,

Z is the impedance of the circuit in ohms,
R is the resistance in ohms,
X_C is the capacitive reactance in ohms.

and,

$$\theta = \tan^{-1}(X_C/R)$$

where,

θ is the phase angle in degrees,
X_C is the capacitive reactance,
R is the resistance.

Suppose you have an RC circuit composed of a 200-Ω resistor connected in series with a capacitor that has a reactance of 500 Ω. What is the impedance of the circuit and what is the phase angle?

From the information stated in the problem, you can see that R = 200 Ω and X_C = 500 Ω. Using the equation for impedance in a series RC circuit:

$$Z = \sqrt{R^2 + X_C^2}$$
$$Z = \sqrt{200^2 + 500^2}$$
$$Z = \sqrt{40{,}000 + 250{,}000}$$
$$Z = \sqrt{290{,}000}$$
$$Z = 539 \; \Omega$$

As far as the phase angle is concerned:

$$\theta = \tan^{-1}(X_C/R)$$
$$\theta = \tan^{-1}(200/500)$$
$$\theta = \tan^{-1}(0.4)$$
$$\theta = 21.8°$$

Total Current in a Series RC Circuit

Many practical situations make it necessary for you to determine the amount of current in a series RC circuit. Ohm's law provides the simplest means for doing this task:

$$I_T = \frac{E_T}{Z}$$

where,

I_T is the total current in amperes,
E_T is the total voltage in volts,
Z is the impedance of the circuit in ohms.

If you apply 120 V ac to a circuit that has an impedance of 500 Ω, this equation shows you that the total current (or current through any of the components) is 120/500, or 0.24 A.

Analyzing Series RC Circuits

If you can analyze series RL circuits as described in Chapter 13, you are ready to analyze series RC circuits. In fact, the only differences between the two procedures are the equations for reactance and the direction of the phase angle.

Figure 16-6 shows a circuit that includes the amplitude and frequency of the ac voltage source, the value of a resistor, and the value of the capacitance of a coil. A complete analysis of this circuit would include the amount of current flowing through the circuit, the reactance of the capacitor, the impedance of the circuit, the voltage drop across each component, and the phase angle between the source voltage and current.

Step 1

Use the given values for frequency and capacitance to calculate the capacitive reactance.

Step 2

Use the given value of R and the value of X_C calculated in Step 1 to determine the total impedance of the circuit.

Step 3

Use the given value of total voltage and the value for Z calculated in Step 2 to determine the total circuit current.

Step 4

Knowing that the current through the resistor is the same as the total current calculated in Step 3, use that value and the given value for R to determine the voltage drop across the resistor.

UNDERSTANDING RC CIRCUITS

**Figure 16-6.
Series RC Circuit
Analysis**

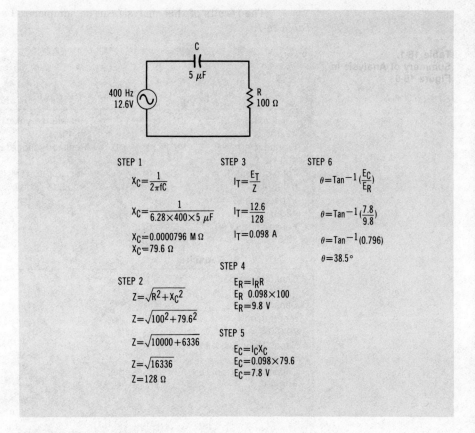

Step 5
Knowing that the capacitive current is the same as the total current calculated earlier in this procedure, use that value and your value for capacitive reactance to determine the voltage drop across the capacitor.

Step 6
Use the calculated values for capacitive and resistive voltage to determine the phase angle.

The results of this analysis can be summarized as shown in *Table 16-1*.

Table 16-1.
Summary of Analysis in Figure 16-6

Circuit		Resistor	
Total voltage: E_T = 12.6 V	(Given)	Resistance value: R = 100 Ω	(Given)
Frequency: f = 400 Hz	(Given)	Resistor voltage: E_R = 9.8 V	(Step 4)
Total current: I_T = 0.098 A	(Step 3)	Resistor current: I_R = 0.098 A	(From I_T; series circuit)
Impedance: Z = 128 Ω	(Step 3)		
Phase shift: Θ = 38.5°	(Step 6)		
Capacitor			
Capacitor value: C = 5 μF	(Given)		
Capacitive reactance: X_C = 79.6 Ω	(Step 1)		
Capacitor voltage: E_C = 7.8 V	(Step 5)		
Capacitive current: I_C = 0.098 A	(From I_T; series circuit)		

PARALLEL RC CIRCUITS

Parallel RC circuits share at least one important characteristic with any other kind of parallel circuit: the voltage is the same across every component in the parallel arrangement. As you can probably guess by now, the total current is less than the sum of the currents through the individual branches.

Voltage in a Parallel RC Circuit

The amplitude and phase of voltages are identical in all parts of a parallel RC circuit.

The same voltage is dropped across every component in a parallel circuit. This means the amplitude of the voltage across every component is equal to the amplitude of the total voltage:

$$E_T = E_C = E_R$$

UNDERSTANDING RC CIRCUITS

where,

E_T is the total voltage in volts,
E_C is the voltage across the capacitance in volts,
E_R is the voltage across the resistance in volts.

The fact that the same voltage appears across every component in a parallel circuit also means that the phase of the voltage is the same for each component.

Current in a Parallel RC Circuit

The amount of current flowing through a resistance is equal to the amplitude of the voltage divided by the value of the resistance: $I_R = E_R/R$. Furthermore, the voltage and current waveforms are exactly in phase, as shown in the resistor waveforms in *Figure 16-7*.

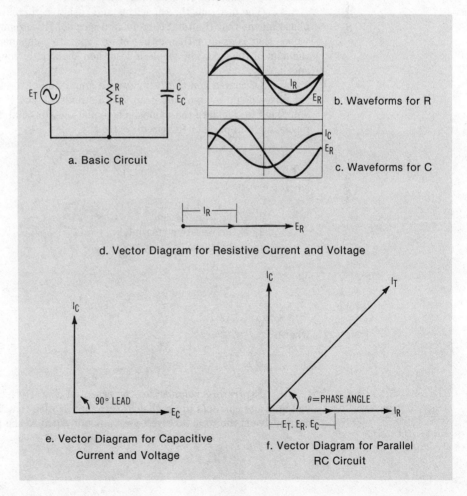

**Figure 16-7.
Currents and Voltages in Parallel RC Circuits**

a. Basic Circuit

b. Waveforms for R

c. Waveforms for C

d. Vector Diagram for Resistive Current and Voltage

e. Vector Diagram for Capacitive Current and Voltage

f. Vector Diagram for Parallel RC Circuit

16 UNDERSTANDING RC CIRCUITS

When working with parallel RC circuits, it is convenient to say that the capacitive current leads the capacitive voltage by 90°.

The current for a capacitance is equal to the amplitude of the voltage dropped across it divided by the value of capacitive reactance: $I_C = E_C/X_C$. However, you already know that the voltage lags the current for a capacitance by exactly 90°. But for the sake of working with parallel RC circuits, it is more meaningful to express the principle in a different way: the current for a capacitor leads the voltage by 90°. This version is shown on the vector diagram for capacitive current and voltage in *Figure 16-7e*.

As usual, the voltages for all components are in phase in a parallel circuit. The currents, however, are 90° out of phase, with the capacitive current leading the resistive current by 90°. Combining the vector diagrams for the resistance and capacitance yields an overall current/voltage vector diagram for parallel RC circuits. As shown in *Figure 16-7f*, the voltages have the same amplitude and phase, the current through the resistor has the same phase as the voltages, and the current for the capacitance leads the resistive current by 90°.

The fact that the capacitive and resistive currents are 90° out of phase means that the total current in a parallel RC circuit is one of those values that is greater than either of its separate components but less than the sum of the two. The whole is less than the sum of the parts, so to speak.

Of course you can estimate the amplitude of the total current and the phase angle by applying graphical methods, or you can use the equations to calculate the values. The equations for total current and the phase angle in parallel RC circuits are:

$$I_T = \sqrt{I_R^2 + I_C^2}$$

where,

I_T is the total current,
I_R is the resistive current,
I_C is the capacitive current,

and,

$$\theta = \mathrm{Tan}^{-1}(I_C/I_R)$$

where,

θ is the phase angle.

The total current in a parallel RC circuit leads the total voltage by some angle less than 90°.

Figure 16-8 summarizes the essential characteristics of the voltages and currents in parallel RC circuits. Notice that the total circuit current leads the total voltage by an amount equal to the phase angle.

UNDERSTANDING RC CIRCUITS

**Figure 16-8.
Current and Voltage
Vector Diagram for
Parallel RC Circuits**

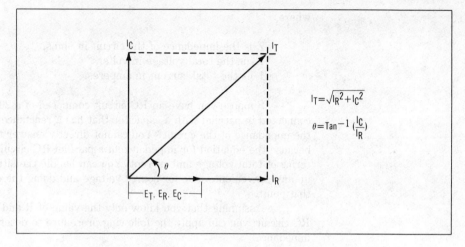

Suppose that the current through the resistance is 3 A and the current through the capacitance is 4 A. Using the equation for the total current in a parallel RC circuit:

$$I_T = \sqrt{I_R^2 + I_C^2}$$
$$I_T = \sqrt{3^2 + 4^2}$$
$$I_T = \sqrt{9 + 16}$$
$$I_T = \sqrt{25}$$
$$I_T = 5 \text{ A}$$

And calculating the phase angle in a parallel RC circuit:

$$\Theta = \text{Tan}^{-1}(I_C/I_R)$$
$$\Theta = \text{Tan}^{-1}(4/3)$$
$$\Theta = \text{Tan}^{-1}(1.33)$$
$$\Theta = 53.1°$$

Impedance of a Parallel RC Circuit

Recall that it is impossible to plot a reliable impedance vector diagram for parallel RL circuits. The same is true for parallel RC circuits, and the roundabout procedure for determining the impedance is quite similar. The basic idea is to use this form of Ohm's law:

$$Z = \frac{E_T}{I_T}$$

where,

> Z is the impedance of the circuit in ohms,
> E_T is the total voltage in volts,
> I_T is the total current in amperes.

Suppose you have an RC circuit composed of a 200-Ω resistor connected in parallel with a capacitor that has a reactance of 500 Ω. What is the impedance of the circuit? You cannot directly answer this question because the equation for impedance of a parallel RC circuit is expressed in terms of total voltage and current. You can handle the situation by making up any convenient value for source voltage and doing the calculations from that point.

Assuming that you know only the value of R and X_C of a parallel RC circuit, you can apply the following procedure to determine the impedance.

Step 1
Select a convenient value for total voltage, E_T.

Step 2
Divide your value of total voltage by the given value of the resistance to determine the current through the resistance.

Step 3
Divide your value for total voltage by the given value of capacitive reactance to determine the amount of capacitive current.

Step 4
Use your values of resistive and capacitive current in the vector equation for total current.

Step 5
Divide your chosen value for total voltage by the total current calculated in Step 4. This provides the value of impedance for the circuit.

By way of a specific example, suppose that you have a parallel RC circuit composed of a 200-Ω resistance and a capacitor that has a capacitive reactance of 500 Ω. Let's apply the procedure just outlined.

1. Pick a convenient voltage such as 100 V.
2. Calculate the current for the resistance:

$$I_R = \frac{E_R}{R}$$

$$I_R = \frac{100}{200}$$

$$I_R = 0.5 \text{ A}$$

3. Use the same voltage to calculate the current for the capacitance:

UNDERSTANDING RC CIRCUITS

$$I_C = \frac{E_C}{X_C}$$

$$I_C = \frac{100}{500}$$

$$I_C = 0.2 \text{ A}$$

4. Use those values of current to calculate the total current in a parallel RC circuit:

$$I_T = \sqrt{I_R^2 + I_C^2}$$
$$I_T = \sqrt{0.5^2 + 0.2^2}$$
$$I_T = \sqrt{0.25 + 0.04}$$
$$I_T = \sqrt{0.29}$$
$$I_T = 0.539 \text{ A}$$

5. Use your value of total current and voltage to calculate the impedance of the circuit:

$$Z = \frac{E_T}{I_T}$$

$$Z = \frac{100}{0.539}$$

$$Z = 186 \text{ }\Omega$$

Analyzing Parallel RC Circuits

Figure 16-9 shows a circuit that specifies the amplitude and frequency of the ac voltage source and the values of resistance and of capacitance. A complete analysis of this circuit includes the voltage drop across each component in the circuit, the reactance of the capacitor, the impedance of the circuit, the current for each component, the total current, and the phase angle between the source voltage and current. The results are summarized for you in *Table 16-2*.

Figure 16-9.
Analysis of Parallel RC Circuit

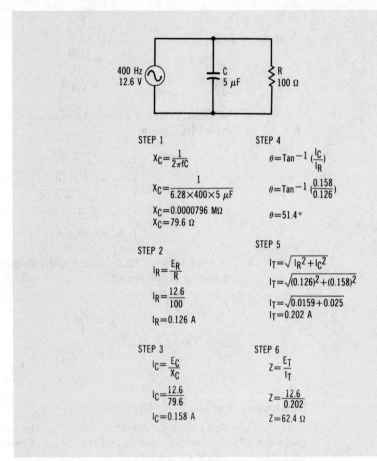

RC CIRCUIT'S EFFECT ON PULSE WAVEFORMS

Most of the features of RC circuits described thus far in this chapter assume that the components are operating from an ac, sine-wave source. This is indeed the most common waveform used with series and parallel RC circuits. However, there are also many instances where pulse waveforms are applied to RC circuits.

UNDERSTANDING RC CIRCUITS — 16

**Table 16-2.
Summary of Analysis in
Figure 16-9**

Circuit		Resistor	
Total voltage: $E_T =$ 12.6 V	(Given)	Resistance value: $R = 100\ \Omega$	(Given)
Frequency: $f =$ 400 Hz	(Given)	Resistor voltage: $E_R = 12.6$ V	(From E_T; parallel circuit)
Total current: $I_T =$ 0.202 A	(Step 5)	Resistor current: $I_R = 0.126$ A	(Step 2)
Impedance: $Z =$ 62.4 Ω	(Step 6)		
Phase shift: $\Theta =$ 51.4°	(Step 4)		

Capacitor	
Capacitance value: $L = 5\ \mu F$	(Given)
Capacitive reactance: $X_C = 79.6\ \Omega$	(Step 1)
Capacitor voltage: $E_C = 12.6$ V	(From E_T; parallel circuit)
Capacitor current: $I_C = 0.158$ A	(Step 3)

When a square-wave pulse is applied to a series RC circuit, the capacitor tends to slow the rise and decay time of the voltage across the capacitor. The first waveform in *Figure 16-10* shows a square waveform applied to a series RC circuit, and the second waveform shows how the capacitance affects the voltage across the capacitance. The third waveform shows the current through the circuit, and the fourth the voltage across the resistance.

**Figure 16-10.
Currents and Voltages
in Pulsed, Series RC
Circuit**

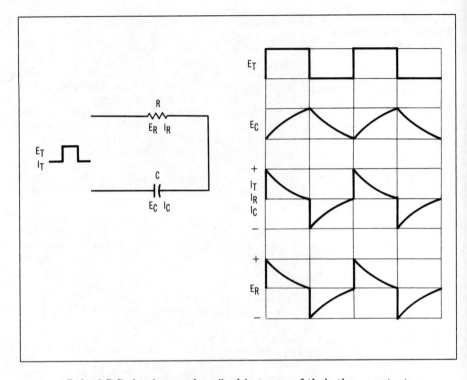

Pulsed RC circuits are described in terms of their time constant. For a series RC circuit, the time constant is defined as the time required for changing the voltage across the capacitor by 63%. The time constant for an RC circuit, in units of seconds, is equal to the value of the resistance in ohms multiplied by the value of capacitance in farads: T = RC.

Suppose that you have a series circuit composed of a 1-MΩ resistor and a 0.01-μF capacitor. What is the time constant of this circuit? From the equation for the time constant of a series RC circuit:

$$T = RC$$
$$T = 1\ \text{M}\Omega \times 0.01\ \mu\text{F}$$
$$T = 1\ \text{second}$$

WHAT HAVE WE LEARNED?

1. An RC circuit is one that includes both a resistance and a capacitance.
2. Currents are equal in amplitude and phase through all parts of a series RC circuit.
3. Current and voltage are in phase through the resistance in a series RC circuit.
4. Voltage lags the current by 90° for the capacitance in a series RC circuit.
5. The total voltage lags the total current for a series RC circuit. The amount of current lag is less than 90°.

UNDERSTANDING RC CIRCUITS 16

6. The total voltage in a series RC circuit is greater than the resistive or capacitive voltages but less than the sum of the two.
7. You can represent the amplitudes and phase relationships in a series RC circuit by means of vector diagrams.
8. You can use graphical methods to determine the total voltage and phase angle in a series RC circuit.
9. Impedance is the opposition to current flow in ac circuits that contain both resistance and reactance. The mathematical symbol for impedance is Z, and the unit of measurement is the ohm.
10. The impedance of a series RC circuit is greater than either the resistance or capacitive reactance, but it is less than the sum of the two.
11. An impedance vector diagram for a series RC circuit shows the relationships between resistance, capacitive reactance, impedance, and the phase angle.
12. Voltages are equal in amplitude and phase across all parts of a parallel RC circuit.
13. Current leads the voltage by 90° for the capacitance in a parallel RC circuit.
14. The total current leads the total voltage for a parallel RC circuit. The amount of current lead is less than 90°.
15. The total current in a parallel RC circuit is greater than the resistive or capacitive currents, but it is less than the sum of the two.
16. You can represent the amplitudes and phase relationships in a parallel RC circuit by means of vector diagrams.
17. You can use graphical methods to determine the total current and phase angle in a parallel RC circuit.
18. There is no such thing as an impedance vector diagram for parallel RC circuits. You can calculate the impedance of a parallel RC circuit, however, by dividing the total voltage by the total current.
19. The time constant of a circuit is the time required for the waveform to change 63% in amplitude.
20. The time constant of a series RC circuit is equal to the value of the resistance multiplied by the capacitance. The result is in units of seconds.

16 UNDERSTANDING RC CIRCUITS

Quiz for Chapter 16

1. Which one of the following statements is true for the voltages in a series RC circuit?
   a. The voltage always has the same amplitude and phase for every part of the circuit.
   b. The total voltage is equal to the sum of the voltages across the resistance and capacitance.
   c. The total voltage is less than the sum of the voltages across the resistance and capacitance.
   d. The total voltage is greater than the sum of the voltages across the resistance and capacitance.
   e. The total voltage leads the total current by less than 90°.

2. Which one of the following statements is true for the currents in a series RC circuit?
   a. The current always has the same amplitude and phase for every part of the circuit.
   b. The total current is equal to the sum of the currents for the resistance and capacitance.
   c. The total current is less than the sum of the currents for the resistance and capacitance.
   d. The total current is greater than the sum of the currents for the resistance and capacitance.
   e. The total current lags the total voltage by less than 90°.

3. Which one of the following statements is true for the impedance of a series RC circuit?
   a. The impedance is equal to the sum of the resistance and capacitive reactance.
   b. The impedance is greater than the sum of the resistance and capacitive reactance.
   c. The impedance is greater than either the resistance or capacitive reactance but less than their sum.
   d. The impedance is less than the sum of the resistance and capacitive reactance.
   e. The impedance cannot be calculated directly.

4. Which one of the following statements is true for the voltage in a parallel RC circuit?
   a. The voltage always has the same amplitude and phase for every part of the circuit.
   b. The total voltage is equal to the sum of the voltages across the resistance and capacitance.
   c. The total voltage is less than the sum of the voltages across the resistance and capacitance.
   d. The total voltage leads the total current by less than 90°.

5. Which one of the following statements is true for the currents in a parallel RC circuit?
   a. The current always has the same amplitude and phase for every part of the circuit.
   b. The total current is equal to the sum of the currents for the resistance and capacitance.

UNDERSTANDING RC CIRCUITS

16

c. The total current is less than the sum of the currents for the resistance and capacitance.
d. The total current is greater than the sum of the currents for the resistance and capacitance.
e. The total current lags the total voltage by less than 90°.

6. Referring to the vector diagrams in *Figure 16-11*, which set best represents the distribution of voltages in a series RC circuit?
   a. Vectors 1A, 1B, and 1C.
   b. Vectors 2A, 2B, and 2C.
   c. Vector 1A only.
   d. Vector 2A only.

**Figure 16-11.
Diagrams for Questions 6–11.**

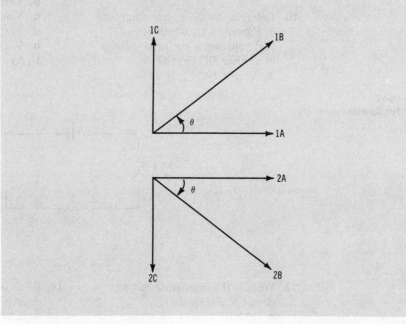

7. Referring to the vector diagrams in *Figure 16-11*, which set best represents the distribution of currents in a parallel RC circuit?
   a. Vectors 1A, 1B, and 1C.
   b. Vectors 2A, 2B, and 2C.
   c. Vector 1A only.
   d. Vector 2A only.

8. Referring to the vector diagrams in *Figure 16-11*, which set best represents an impedance diagram for a series RC circuit?
   a. The set made up of vectors 1A, 1B, and 1C.
   b. The set made up of vectors 2A, 2B, and 2C.
   c. Series impedance cannot be represented by a vector diagram.

16 UNDERSTANDING RC CIRCUITS

9. Referring to the vector diagrams in *Figure 16-11*, which set best represents an impedance diagram for a parallel RC circuit?
   a. The set made up of vectors 1A, 1B, and 1C.
   b. The set made up of vectors 2A, 2B, and 2C.
   c. Parallel impedance cannot be represented by a vector diagram.

10. Referring to the vector diagrams in *Figure 16-11*, which vector best represents the total voltage in a series RC circuit?
    a. Vector 1A.
    b. Vector 1B.
    c. Vector 1C.
    d. Vector 2A.
    e. Vector 2B.
    f. Vector 2C.

11. Referring to the vector diagrams in *Figure 16-11*, which vector best represents the total current in a parallel RC circuit?
    a. Vector 1A.
    b. Vector 1B.
    c. Vector 1C.
    d. Vector 2A.
    e. Vector 2B.
    f. Vector 2C.

Figure 16-12.
Circuit for Questions 12-15

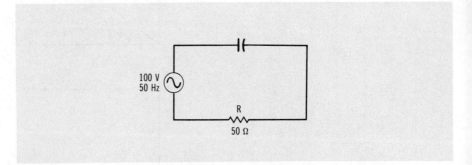

12. What is the impedance of the circuit in *Figure 16-12*?
    a. 19.3 Ω.
    b. 31.8 Ω.
    c. 40.7 Ω.
    d. 50.1 Ω.
    e. 59.2 Ω.
    f. 81.4 Ω.

13. What is the phase angle between total current and voltage for the the circuit in *Figure 16-12*?
    a. 0°.
    b. 32.4°.
    c. 57.6°.
    d. 90°.

14. What is the total current for the circuit in *Figure 16-12*?
    a. 1.23 A.
    b. 1.69 A.
    c. 2 A.
    d. 2.46 A.
    e. 5.8 A.

15. What is the voltage across the resistor in *Figure 16-12*?
    a. 46.9 V.
    b. 53.1 V.
    c. 61.5 V.
    d. 84.5 V.
    e. 100 V.
    f. 123 V.

UNDERSTANDING RC CIRCUITS

16. What is the impedance of a circuit composed of a 100-Ω resistor connected in parallel with a capacitor that has a reactance of 100 Ω?
    a. 50 Ω.
    b. 89.3 Ω.
    c. 150 Ω.
    d. 200 Ω.
    e. 300 Ω.
    f. It cannot be determined from the information supplied.

17. What is the phase shift between total current and voltage in the circuit described in Question 16?
    a. 0°.
    b. 26.6°.
    c. 45°.
    d. 63.4°.
    e. 90°.
    f. It cannot be determined from the information supplied.

17

Understanding RLC Circuits

ABOUT THIS CHAPTER

This chapter gives you a chance to combine everything you know about resistive, inductive, and capacitive circuits. You will learn how combinations of resistors, inductors, and capacitors can be combined to create the effects of series and parallel resonance. When you complete this chapter you should be able to identify resonant circuits, describe their behavior as the applied frequency changes, and calculate the total impedance of these circuits.

RLC CIRCUITS?

An RLC circuit is any circuit that includes a resistance, an inductance, and a capacitance. *Figure 17-1* shows the two basic kinds of RLC circuits—one where the components are connected in series and another where they are connected in parallel. Of course there are many other possible combinations of these three basic components and a great many other combinations with two or more components of the same kind. But if you understand the fundamental principles of the two basic RLC circuits, you will be able to determine the characteristics of variations when the need arises.

17 UNDERSTANDING RLC CIRCUITS

**Figure 17-1.
RLC Circuits**

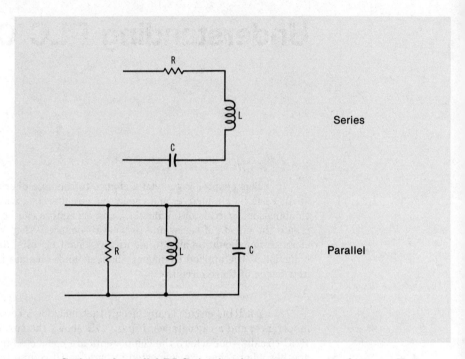

Series and parallel RLC circuits share some common features. You should be aware of those common features before you attempt to understand other important characteristics in greater detail.

How Frequency Affects the Impedance of R, L, and C Devices

Resistance in an RLC circuit remains unchanged as the applied frequency changes.

In any kind of circuit, the value of a resistance remains unchanged as the frequency of the circuit changes. As shown in *Figure 17-2*, though, changing the frequency applied to inductive and capacitive circuits has a great effect on their reactance.

Inductive Reactance

The inductive reactance in a circuit is determined by this equation:

$$X_L = 2\pi f L$$

where,

X_L is the amount of inductive reactance in ohms,
π is the constant pi, 3.14,
f is the applied frequency in hertz,
L is the inductance in henrys.

UNDERSTANDING RLC CIRCUITS   17

**Figure 17-2.
Effects of Frequency on
Resistance, Inductive
Reactance, and
Capacitive Reactance.**

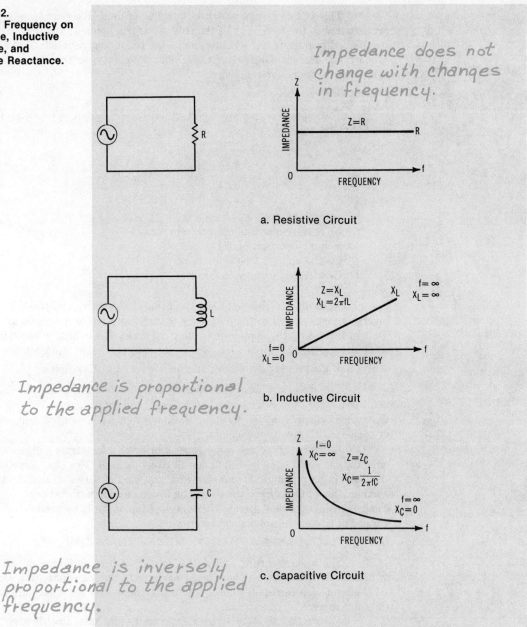

a. Resistive Circuit

b. Inductive Circuit

c. Capacitive Circuit

17 UNDERSTANDING RLC CIRCUITS

The value of inductive reactance in an RLC circuit is proportional to the applied frequency; as the applied frequency increases, the inductive reactance increases a proportional amount.

This equation suggests that the value of inductive reactance is proportional to the frequency of the sine waveform applied to it. Increasing the applied frequency, for instance, increases the inductive reactance by a proportional amount. Decreasing the applied frequency, on the other hand, decreases the inductive reactance.

Capacitive Reactance

The capacitive reactance of a circuit follows a somewhat different kind of equation:

$$X_C = \frac{1}{(2\pi fC)}$$

where,

X_C is the capacitive reactance in ohms,
π is the constant pi, 3.14,
f is the applied frequency in hertz,
C is the capacitance in farads.

The value of capacitive reactance in an RLC circuit is inversely proportional to the applied frequency; as the applied frequency increases, the capacitive reactance decreases a proportional amount.

According to this equation, the value of capacitive reactance is inversely proportional to the frequency of the sine waveform applied to it. That is, increasing the applied frequency decreases the inductive reactance, while decreasing the applied frequency increases the inductive reactance.

Voltage, Current, and Impedance for RLC Circuits

You have just seen how changes in frequency affect the impedance and current for pure resistances, inductances, and capacitances. We will see that these principles apply, whether the components are connected in series or parallel with other components.

Current has the same amplitude and phase through every component in a series circuit. Voltage has the same amplitude and phase across every branch in a parallel circuit.

Likewise, there are some general principles for series and parallel circuits that apply equally well under all circumstances. For one, current is the same in both amplitude and phase for any kind of series circuit. And for another, the voltages have the same amplitude and phase for every component in a parallel circuit. These are principles we have used in a variety of circuits in previous chapters.

HOW FREQUENCY AFFECTS SERIES RLC CIRCUITS

The basic nature of series RLC circuits is best described in terms of impedance. Once you see how frequency affects the impedance of a series RLC circuit, you can apply Ohm's law to determine how frequency affects the total current for the circuit.

Figure 17-3 shows a series RLC circuit. You have already seen how frequency affects pure resistances, inductive reactances, and capacitive reactances separately. Combining these three effects in a series RLC creates an entirely different picture.

**Figure 17-3.
Series RLC Circuit**

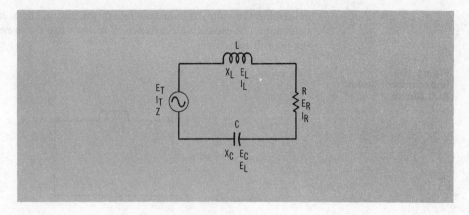

Impedance of a Series RLC Circuit

The total impedance of a series RLC circuit is the overall opposition to current flow offered by the resistance, inductive reactance, and capacitive reactance. Because the values of inductive and capacitive reactance change as the applied frequency changes, it follows that the impedance changes accordingly.

The impedance of a series RLC circuit is determined by this equation:

$$Z = \sqrt{R^2 + (X_L - X_C)^2}$$

where,

Z is the impedance of the circuit in ohms,
R is the value of the resistance in ohms,
X_L is the reactance of the inductor in ohms,
X_C is the reactance of the capacitor in ohms.

If the resistance in a certain series RLC circuit is 100 Ω, the inductive reactance is 200 Ω, and the capacitive reactance is 150 Ω, you can use the formula to determine the impedance.

$$Z = \sqrt{R^2 + (X_L - X_C)^2}$$
$$Z = \sqrt{100^2 + (200 - 150)^2}$$
$$Z = \sqrt{100^2 + 50^2}$$
$$Z = \sqrt{12{,}500}$$
$$Z = 112 \; \Omega$$

When you are working with practical RLC circuits, you rarely know the values of inductive and capacitive reactance. So you have to calculate those values before you can determine the impedance of the circuit. *Figure 17-4* illustrates this situation. In the first step, you determine the value of inductive reactance, and in the second, you determine the capacitive reactance. You then use these results to calculate the overall impedance.

Figure 17-4. Impedance of Series RLC Circuit

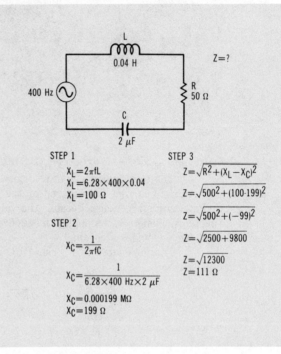

When the inductive reactance is greater than the capacitive reactance in a series RLC circuit, the circuit behaves as though it is an RL circuit. The capacitive reactance cancels out some of the inductive reactance, but the remaining inductive reactance is sufficient to display the essential characteristics of an inductive circuit: the current lags the applied voltage, for example.

When the capacitive reactance in a series RLC circuit is greater than the inductive reactance, the difference between the two leaves some amount of capacitive reactance. As a result, the circuit behaves as though it is an RC circuit. The current in such a circuit leads the applied voltage.

Resonant Frequency of a Series RLC Circuit

The impedance of a series RLC circuit is equal to the value of the resistance when the inductive and capacitive reactances are equal.

Something very special happens when the inductive reactance is equal to the capacitive reactance. At this particular point, the difference between the two values is zero, so the impedance is equal to the value of the resistance alone. The circuit still has inductive and capacitive reactance, but the two cancel each other's effects. The overall impression is that the circuit is neither inductive nor capacitive but purely resistive.

UNDERSTANDING RLC CIRCUITS

The frequency where inductive and capacitive reactances are equal is called the circuit's resonant frequency.

In an RLC circuit, the frequency at which the impedance is equal to the value of the resistance (where the inductive and capacitive reactances are exactly equal) is called the "resonant frequency." An RLC circuit has just one resonant frequency. You can determine the value of a circuit's resonant frequency with this equation:

$$f_o = \frac{1}{(2\pi\sqrt{LC})}$$

where,

f_o is the resonant frequency in hertz,
π is the constant pi, 3.14,
L in the value of the inductor in henrys,
C is the value of the capacitor in farads.

When the frequency applied to a series RLC circuit is equal to the resonant frequency, the current is in phase with the total voltage and the circuit is purely resistive.

When a series RLC circuit is operated below its resonant frequency, the inductive reactance dominates the behavior of the circuit, and the current lags the total voltage by a phase angle that is less than 90°. When the applied frequency is above the resonant frequency, the capacitive effect dominates the activity, and the current leads the applied voltage by some angle less than 90°. When the applied frequency is equal to the circuit's resonant frequency, the circuit is neither inductive nor capacitive, and the current is in phase with the total voltage.

Total Current in a Series RLC Circuit

The simplest and most reliable way to determine the total current in a series RLC circuit is to apply this form of Ohm's law:

$$I_T = \frac{E_T}{Z}$$

where,

I_T is the circuit current in amperes,
E_T is the total voltage in volts,
Z is the total impedance in ohms.

If the total impedance of a series RLC circuit is 10 Ω and the applied voltage is 60 V, you can be certain that the circuit current is 60/10 or 6 A.

17 UNDERSTANDING RLC CIRCUITS

Table 17-1 summarizes the ways a series RLC circuit responds to frequencies above, below, and at the resonant frequency.

Table 17-1. Summary of Responses of Series RLC Circuits

Below Resonance	At Resonance	Above Resonance
1. Circuit is capacitive	1. Circuit is resistive	1. Circuit is inductive
2. Source current leads the source voltage	2. Source current and voltage are in phase	2. Source current lags the source voltage
3. Current is less than maximum	3. Current is maximum	3. Current is less than maximum
4. Impedance is greater than R	4. Impedance is minimum and equal to R	4. Impedance is greater than R

HOW FREQUENCY AFFECTS PARALLEL RLC CIRCUITS

The essential character of parallel RLC circuits is best described in terms of currents in the individual branches. Once you see how frequency affects the current in the individual branches, you can figure the total current and apply Ohm's law to determine how frequency affects the total impedance of the circuit.

Figure 17-5 shows a parallel RLC circuit. You have already seen how frequency affects the current through pure resistances, inductive reactances, and capacitive reactances separately. You have also seen how frequency affects series RLC circuits. The behavior of a parallel RLC circuit requires that you think through the same process, but you will see that the effects are quite different.

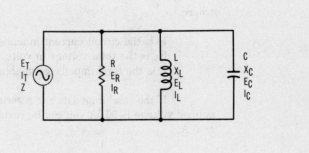

Figure 17-5. Parallel RLC Circuit

Currents in a Parallel RLC Circuit

The total current of a parallel RLC circuit is determined by this equation:

UNDERSTANDING RLC CIRCUITS

$$I_T = \sqrt{I_R^2 + (I_C - I_L)^2}$$

where,

I_T is the total circuit current,
I_R is the current through the resistive branch,
I_C is the current through the capacitive branch,
I_L is the current through the inductive branch.

If the resistive current in a parallel RLC circuit is 0.1 A, the capacitive current is 2 A, and the inductive current is 1.5 A, you can use the equation to determine the total current.

$$I_T = \sqrt{0.1^2 + (2 - 1.5)^2}$$
$$I_T = \sqrt{0.1^2 + 0.5^2}$$
$$I_T = \sqrt{0.01 + 0.25}$$
$$I_T = \sqrt{0.26}$$
$$I_T = 0.51 \text{ A}$$

When you are working with practical RLC circuits, you rarely know the values of the current in each branch. This means you must be prepared to carry out some preliminary steps. First, you have to calculate the capacitive and inductive reactances on the basis of the values of those components and the frequency applied to the circuit. You can then apply Ohm's law to each component to determine the current for each branch based on the voltage applied to them. *Figure 17-6* illustrates this procedure in detail.

When the inductive current is greater than the capacitive current in a parallel RLC circuit, the circuit behaves as though it is an RL circuit. The capacitive current cancels out some of the effects of the inductive current, but the remaining inductive current is sufficient to display the essential characteristics of an inductive circuit: the total current lags the applied voltage, for example.

When the capacitive current in a parallel RLC circuit is greater than the inductive current, the difference between the two leaves some amount of capacitive current. As a result, the circuit behaves as though it is an RC circuit. The total current in such a circuit leads the applied voltage.

**Figure 17-6.
Current in Parallel RLC Circuit**

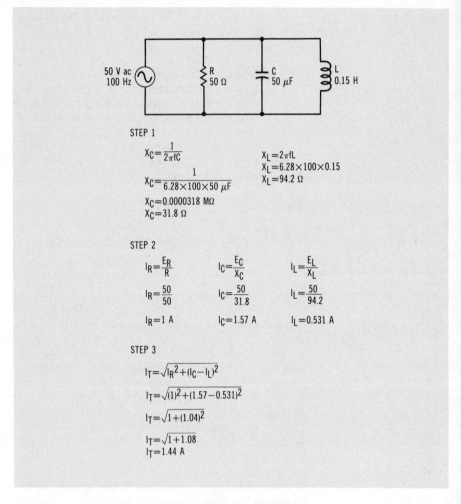

Resonant Frequency of a Parallel RLC Circuit

When a parallel RLC circuit is operating at its resonant frequency, the capacitive and inductive currents are equal; the total current is equal to the current for the resistive branch of the circuit.

The resonant frequency of a parallel RLC circuit is the frequency where the inductive and capacitive currents are equal. When the capacitive current equals the inductive current, the equation for total current shows that the total current is equal to the resistive current. In theory at least, LC components in a parallel circuit do not use any source current at the circuit's resonant frequency.

The equation for determining the resonant frequency for a parallel RLC circuit is the same as the one used for this purpose for series RLC circuits: $f_o = 1/(2\pi\sqrt{LC})$.

UNDERSTANDING RLC CIRCUITS

When the frequency applied to a series RLC circuit is equal to the resonant frequency, the current is in phase with the total voltage and the circuit is purely resistive.

When a parallel RLC circuit is operated below its resonant frequency, the inductive current dominates the behavior of the circuit, and the current lags the applied voltage by a phase angle that is less than 90°. When the applied frequency is above the resonant frequency, the capacitive effect dominates the activity, and the current leads the applied voltage by some angle less than 90°. When the applied frequency is equal to the circuit's resonant frequency, the circuit is neither inductive nor capacitive, and the current is in phase with the applied voltage.

Impedance of a Parallel RLC Circuit

The simplest way to determine the total impedance of a parallel RLC circuit is by applying Ohm's law in terms of the calculated value of total current and the applied voltage: $Z = E_T/I_T$.

Table 17-2 summarizes the behavior of a parallel RLC circuit in terms of frequencies that are below, above, and at the resonant frequency.

Table 17-2.
Summary of Responses of Parallel RLC Circuits

Below Resonance	At Resonance	Above Resonance
1. Circuit is inductive	1. Circuit is resistive	1. Circuit is capacitive
2. Source current lags the source voltage	2. Source current and voltage are in phase	2. Source current leads the source voltage
3. Current is greater than minimum	3. Current is minimum	3. Current is greater than minimum
4. Impedance is less than R	4. Impedance is maximum and equal to R	4. Impedance is less than R

Q FACTOR OF AN RLC CIRCUIT

We have seen that practical inductors include a bit of wire resistance as well as inductance. The quality, or Q factor, of these inductors is equal to the ratio of their inductive reactance to their resistance: $Q = X_L/R$. The higher the Q factor, the more the device behaves as a pure inductor.

RLC circuits also have Q factors, but they have to be defined in terms of capacitance, inductance, and resistance. The Q factor of an RLC circuit is an indication of how well the circuit responds to changes in frequency near the resonant frequency. The higher the Q factor, the more the response matches the response of a pure LC circuit.

The Q factor indicates the sharpness of a circuit's frequency curve. As shown in *Figure 17-7*, the higher the Q factor, the sharper and higher the response. Low-Q circuits show less distinct changes over a wider range of frequencies.

**Figure 17-7.
Q Factor Affects
Frequency Response**

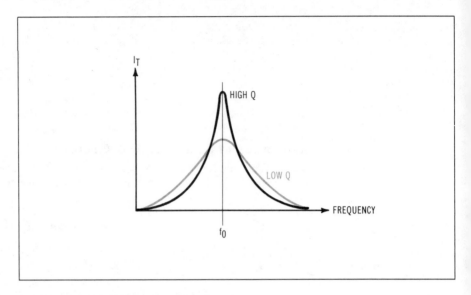

WHAT HAVE WE LEARNED?
1. An RLC circuit contains resistance, inductance, and capacitance.
2. Resistance is not affected by changes in applied frequency.
3. Inductive reactance increases as the applied frequency increases.
4. Capacitive reactance decreases as the applied frequency increases.
5. Series RLC circuits are best described in terms of total impedance.
6. The resonant frequency of an RLC circuit is the frequency where the inductive reactance is equal to the capacitive reactance.
7. The impedance of a series RLC circuit is at a minimum (equal to the value of R) at resonance.
8. A series RLC circuit is capacitive when operated below its resonant frequency, it is purely resistive at the resonant frequency, and it is inductive when operated above its resonant frequency.
9. Parallel RLC circuits are best described in terms of current for each of the branches.
10. The impedance of a parallel RLC circuit is at a maximum when it is operated at its resonant frequency.
11. A parallel RLC circuit is inductive when operated below its resonant frequency, it is purely resistive at the resonant frequency, and it is capacitive when operated above its resonant frequency.
12. The Q factor of an RLC circuit is an indicator of the sharpness of a circuit's frequency curve: the higher the Q factor, the sharper and higher the response.

KEY WORD
Resonant frequency

Quiz for Chapter 17

1. Which one of the following statements is true?
   a. Resistance increases as the applied frequency increases.
   b. Resistance decreases as the applied frequency increases.
   c. Resistance is not affected by changes in frequency.

2. Which one of the following statements is true?
   a. Inductive reactance increases as the applied frequency increases.
   b. Inductive reactance decreases as the applied frequency increases.
   c. Inductive reactance is not affected by changes in frequency.

3. Which one of the following statements is true?
   a. Capacitive reactance increases as the applied frequency increases.
   b. Capacitive reactance decreases as the applied frequency increases.
   c. Capacitive reactance is not affected by changes in frequency.

4. Which one of the following statements best applies to a parallel RLC circuit?
   a. The voltage waveform for each component always has the same amplitude and phase as the applied voltage.
   b. The current waveform for each component always has the same amplitude and phase as the applied current.
   c. The sum of the currents is always less than the applied current.

5. Which one of the following statements best applies to a series RLC circuit?
   a. The voltage waveform for each component always has the same amplitude and phase as the applied voltage.
   b. The current waveform for each component always has the same amplitude and phase as the applied current.
   c. The sum of the currents is always less than the applied current.

6. The total impedance of a series RLC circuit:
   a. is equal to the sum of the values of resistance, inductive reactance, and capacitive reactance.
   b. always increases as the applied frequency increases.
   c. always decreases as the applied frequency increases.
   d. is maximum at the resonant frequency.
   e. is minimum at the resonant frequency.

7. When a series RLC circuit is operating at its resonant frequency:
   a. inductive reactance is greater than the capacitive reactance.
   b. capacitive reactance is greater than the inductive reactance.
   c. inductive reactance is equal to the capacitive reactance.
   d. the difference between inductive and capacitive reactance is equal to the resistance.

8. When operating below its resonant frequency, a series RLC circuit has the characteristics of a:
   a. series RL circuit.
   b. series RC circuit.
   c. purely resistive circuit.
   d. series RLC circuit operated above its resonant frequency.

9. When operating above its resonant frequency, a series RLC circuit has the characteristics of a:
   a. series RL circuit.
   b. series RC circuit.
   c. purely resistive circuit.
   d. series RLC circuit operated below its resonant frequency.

10. The total impedance of a parallel RLC circuit:
    a. is equal to the sum of the values of resistance, inductive reactance, and capacitive reactance.
    b. always increases as the applied frequency increases.
    c. always decreases as the applied frequency increases.
    d. is maximum at the resonant frequency.
    e. is minimum at the resonant frequency.

11. When a parallel RLC circuit is operating at its resonant frequency:
    a. inductive reactance is greater than the capacitive reactance.
    b. capacitive reactance is greater than the inductive reactance.
    c. inductive reactance is equal to the capacitive reactance.
    d. the difference between inductive and capacitive reactance is equal to the resistance.

12. When operating below its resonant frequency, a parallel RLC circuit has the characteristics of a:
    a. parallel RL circuit.
    b. parallel RC circuit.
    c. purely resistive circuit.
    d. parallel RLC circuit operated above its resonant frequency.

13. When operating above its resonant frequency, a parallel RLC circuit has the characteristics of a:
    a. parallel RL circuit.
    b. parallel RC circuit.
    c. purely resistive circuit.
    d. parallel RLC circuit operated below its resonant frequency.

Understanding Transformers

You are about to use your understanding of ac electricity to learn about transformers. You will learn how a transformer transfers power from one winding to another. You will learn how to calculate the changes in voltage, current, and impedance produced by a transformer with a known turns ratio. You will also see how you can adjust the turns ratio to get the transformer effects you need.

TRANSFORMER DEFINED

A "transformer" is an inductive device that transforms certain electrical quantities from one level to another. Some transformers are used to transform ac voltage and currents from one level to another. Other transformers are used to match the impedance of one circuit to the impedance of another.

A basic transformer has a primary winding and a secondary winding wound on the same core.

A basic transformer has two different windings—a primary and a secondary winding—wound on the same core. The core material can be laminated iron, molded powdered iron (ferrite core), or air.

Ac current applied to the primary winding of a transformer induces a voltage in the secondary winding.

Through the principles of induction, alternating current flowing in the primary winding of a transformer sets off an alternating magnetic field in the core. This magnetic field, in turn, induces an alternating voltage in the secondary winding. In this way, a transformer transfers energy from the primary to the secondary winding. Ac power is applied to the primary winding of a transformer, and the "transformed" power is taken from the secondary winding.

Power taken from the secondary winding of a transformer cannot be greater than the power applied to the primary.

One of the most important principles of transformers is that the power derived from the secondary winding cannot be greater than the power applied to the primary winding. As we will see, the secondary power is always slightly less than the power applied to the primary.

Tranformers can be classified as power transformers or impedance-matching transformers. "Power transformers" are designed to operate at a single frequency, and they alter current and voltage levels in order to supply levels required for operating electrical equipment and electronic devices. "Impedance-matching transformers" are used where it is necessary to match the impedance of one electrical circuit to the impedance of another. Impedance-matching transformers are usually designed to operate over a wide range of ac frequencies.

18 UNDERSTANDING TRANSFORMERS

PRINCIPLES OF POWER TRANSFORMERS

The main purpose of a power transformer is to transform voltage and current levels. *Figure 18-1* illustrates the construction of a basic power transformer and shows its schematic symbol. The primary and secondary windings are wrapped around a laminated iron core. You connect the primary winding to the source of power and connect the secondary winding to the electrical device to be operated. As alternating current flows through the primary, an alternating magnetic field cuts the turns of the secondary winding, thereby generating a voltage in the secondary winding.

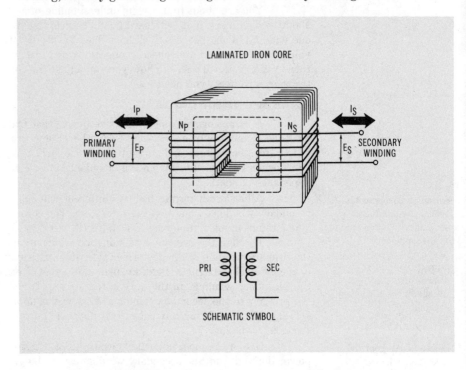

Figure 18-1.
Basic Power Transformer

Power transformers are designed to operate from ac sine waves at one particular frequency.

Power transformers are designed to operate with sine-wave ac power. Furthermore, each is designed to operate at a certain frequency. The most common operating frequency is 60 Hz—the frequency of U.S. power utility companies. Some power transformers are designed for use at 50 Hz, which is the standard utility frequency for many other parts of the world. There are also power transformers that operate at the 400-Hz frequency, which is often used in aerospace and military equipment.

Most of the principles and equations we will examine apply to transformers that are operating from sine waves. You cannot expect the same results when using a transformer with other wave shapes.

UNDERSTANDING TRANSFORMERS

A step-down transformer steps down a voltage level. The secondary voltage of a step-down transformer is always less than the voltage applied to its primary winding.

A transformer that reduces the voltage in a circuit is called a "step-down transformer." The voltage taken from the secondary winding of a step-down transformer is always less than the voltage applied to the primary winding. You have seen many examples in this book that cite ac voltages of 12.6 V. This represents a very popular voltage rating for filament transformers. A filament transformer is a step-down transformer that usually transforms 120 V down to 12.6 V. You can also buy filament transformers that step down 120 V to 6.3 V.

A step-up transformer steps up a voltage level. The secondary voltage of a step-up transformer is always greater than the voltage applied to its primary winding.

A transformer that steps up a voltage level is called a "step-up transformer." The secondary voltage from a step-up transformer is always greater than the voltage applied to its primary winding. A neon sign transformer is a common example of a step-up transformer. Neon signs require operating voltages on the order of 12,000 volts. These transformers step up the commonly available 120 V ac to 12,000 V ac.

Turns Ratio of a Transformer

The turns ratio of a transformer is the ratio of the number of turns of wire in the primary winding to the number of turns in the secondary winding.

The amount of voltage induced in the secondary winding of a transformer depends on how many turns of wire are in the primary compared with the secondary winding. The number of turns of wire in the primary divided by the number of turns in the secondary is called the "turns ratio." You can express the turns ratio by means of a mathematical statement:

$$\text{turns ratio} = \frac{N_p}{N_s}$$

where N_p is the number of turns in the primary winding and N_s is the number of turns in the secondary winding.

Voltages for a Power Transformer

Assuming that you are operating the transformer from a sinusoidal power source and at its rated frequency, you can use the following equation to express the relationship between voltages and turns ratio.

$$\frac{E_p}{E_s} = \frac{N_p}{N_s}$$

where,

E_p is the primary voltage in volts,
E_s is the secondary voltage in volts,
N_p/N_s is the turns ratio (no units),
E_p is the primary voltage in volts.

Consider a power tranformer that has a turns ratio of 100. As long as you know this figure, you do not have to know the exact number of turns in each case. The ratio is just as useful. Further suppose that you apply 120 V to the primary of this transformer. What is the secondary voltage? According to the voltage equation:

$$\frac{Ep}{Es} = \frac{Np}{Ns}$$

$$\frac{120}{Es} = 100$$

$$Es = 1.2 \text{ V}$$

A step-up power transformer has a turns ratio that is less than 1. A step-down transfomer has a turns ratio greater than 1. An isolation transformer has a turns ratio of 1.

The transformer in this example steps down the voltage applied to the primary winding. It is thus a step-down transformer. Notice that the turns ratio for a step-down transformer is a value greater than 1. A step-up tranformer must have a turns ratio that is less than 1. Consider, for example, the secondary voltages from a transformer that has a turns ratio of 0.1. A power transformer that has a turns ratio equal to 1—having the same number of turns in the primary and secondary windings—is called an "isolation transformer."

An isolation transformer does not alter the voltage and current levels. It is used in equipment that has to be "isolated" from the utility power system for safety purposes. All power transformers that have separate primary and secondary windings perform this isolation function, but only those having a turns ratio of 1 are called isolation transformers.

Currents for Power Transformers

You have already learned that the power taken from the secondary cannot exceed the amount of power being applied to the primary winding. This means that you cannot step up a voltage without reducing the current capability by a proportional amount. In an ideal power transformer (one where the secondary power is equal to the primary power), you can use the following expression of transformer current levels:

$$\frac{Is}{Ip} = \frac{Np}{Ns}$$

where,

Is is the secondary current in amperes,
Ip is the primary current in amperes,
Np/Ns is the turns ratio (no units).

UNDERSTANDING TRANSFORMERS   18

You can determine the turns ratio of a power transformer by applying a known ac voltage to the primary winding and measuring the voltage at the secondary winding. The ratio, Ep/Es, is the same as the turns ratio.

Knowing the turns ratio of a transformer makes it possible for you to predict the amounts of voltage and current you can expect to find at the primary and secondary windings. In the case of power transformers, however, you rarely find a specification for their turns ratio. All you know is the maximum voltage rating of the primary winding. But you can determine these characteristics of a power transformer by applying a known ac voltage to the primary winding and measuring the secondary voltage with an ac voltmeter. The ratio of those two voltages, Ep/Es, is equal to the turns ratio.

Another equation that can be a great practical value is:

$$\frac{E_p}{E_s} = \frac{I_s}{I_p}$$

Figure 18-2 illustrates a practical example. Suppose you are working with a filament transformer that produces a 12 V ac voltage at its secondary winding when you apply 120 V ac to the primary. You also know that the maximum secondary current is going to be 5 A. What is the primary current under this condition?

$$\frac{E_p}{E_s} = \frac{I_s}{I_p}$$

$$\frac{120}{12} = \frac{5}{I_p}$$

$$10 = \frac{5}{I_p}$$

$$I_p = 0.5 \text{ A}$$

The maximum primary current is going to be 0.5 A in this example. You can thus protect the circuit from a short-circuit condition by using a 0.5-A fuse in the primary winding.

**Figure 18-2.
Selecting a Fuse for a Power Transformer Circuit**

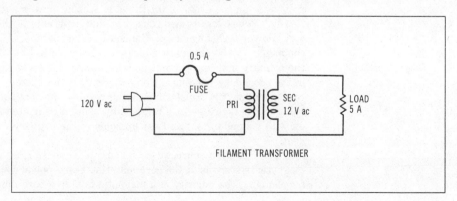

Power Transformer Ratings

Power transformers are rated according to their primary and secondary voltage, operating frequency, and maximum operating power. By knowing the primary and secondary voltage ratings, you can calculate the current ratings and turns ratio. If the specifications for the transformer do not include the operating frequency, you can assume that it is 60 Hz. The power rating is especially important and requires more explanation.

The power rating of transformers is expressed in watts or volt-amperes.

Power ratings for power transformers are expressed in terms of watts or volt-amperes (the product of volts times amperes and abbreviated VA). These voltage and power ratings give you a very good idea about how much current the transformer can handle.

Imagine that you need a power transformer that steps 120 V ac down to 12.6 V ac, and you want the secondary winding to be able to supply 1.5 A of current. What is the minimum power rating of the transformer required for this task?

You know that the secondary winding has to supply 12.6 V at 1.5 A. Power is equal to voltage multiplied by current, so the power demand for the secondary winding is $12.6 \times 1.5 = 18.9$ W. You cannot find a power transformer rated at exactly 18.9 watts or volt-amperes, so you have to select one with a slightly greater power rating—20 W, for example.

Transformer Efficiency and Losses

Practical power transformers operate at less than 100% efficiency.

So far we have assumed that no power is lost in the transfer between the primary and secondary windings. However, no transformer operates at 100% efficiency—the power at the secondary is always going to be a bit less than the power supplied at the primary winding. As a result, secondary voltages are a bit lower than the basic equations would have us believe.

The power loss between the primary and secondary windings is due to a combination of losses. For one, there is a slight power loss caused by the resistance of the wiring itself. If you carefully measure the resistance of the primary winding in a power transformer, you might find a resistance on the order of 0.1 Ω. At 10 A, that means a power loss of 0.1 W.

Power loss is also caused by eddy currents in the transformer core. Power transformers use laminated iron cores. Iron is a fairly good conductor, so the alternating magnetic fields caused by current applied to the primary induce a current flow in a conductive core material as well as in the secondary winding. "Eddy current" is the current induced in the core material of a transformer as the result of normal transformer action. Eddy currents are greatly reduced by using a laminated core, but they still exist to a small extent and use up some of the primary-winding power as heat.

The third source of power loss in a transformer is caused by the magnetic hysteresis of the core material. Some power is always required for changing the polarity of a magnetic field in a core material.

UNDERSTANDING TRANSFORMERS

18

Power transformers may run at a temperature that is hot to the touch, but not so hot that they begin to burn.

Wire resistance, eddy currents, and hysteresis losses all contribute to unwanted power losses in transformers. They account for the fact that power transformers tend to run hotter than most other components in modern electronic systems. If you can press your fingers against an operating power transformer without having to grit your teeth in pain, it is probably operating within its specified power limits. You know you are in trouble, though, when you see blue smoke and smell burning wire.

Power transformers have power efficiencies of 97% or better.

Fortunately, most commercial grade transformers operate at 97% efficiency or better. This means the general equations for calculating voltage, current, and power levels are accurate within 3% and that is a reasonable tolerance for most electronic applications.

Center-Tapped Transformers

Figure 18-3 shows a popular type of power transformer called a "center-tapped transformer." This example has a center-tapped secondary winding. The "tap" is nothing more than an electrical connection to the middle winding of the secondary.

**Figure 18-3.
A Center-Tapped Power Transformer**

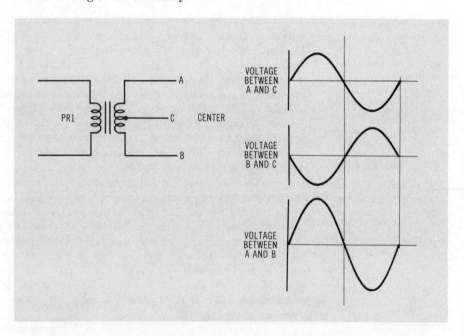

When working with a transformer that has a center-tapped secondary, you will find the full secondary voltage between the two ends of the winding—between points A and B in *Figure 18-3*. However you will find one-half the full secondary voltage between the center tap and either end of the winding. The waveforms in the diagram show 6.3 V ac measured between points A and C and between points B and C. The voltage between points A and B is twice that amount—12.6 V.

UNDERSTANDING ELECTRICITY AND ELECTRONICS CIRCUITS

Power transformers with center-tapped secondary windings are used in the most common kinds of power supplies (units that convert ac to dc) and for getting two or more different ac voltage levels. Some power transformers have center-tapped primary windings, which makes them adaptable for use with either the standard 120-V ac power lines or the 24-V ac version.

PRINCIPLES OF IMPEDANCE-MATCHING TRANSFORMERS

Most electrical and electronic equipment performs more than one operation: input and output sections at least. There is usually an input section that transforms one kind of energy to electrical energy. The input to a public address system, for example, is a microphone that transforms sound into electrical energy. A computer keyboard transforms key depressions into electrical pulses. Also, there is usually an output section that transforms electrical energy into a different form of energy. An audio system, for instance, uses loudspeakers to transform electrical energy into sound. And there may be any number of sections between the input and output sections.

Impedance-matching transformers are used for matching the output impedance of one part of a circuit to the input impedance of the next.

So you can view a piece of electronic equipment as a series of sections, with each section manipulating the signals in some useful fashion. The problem is that the output impedance of one section should match the input section of the next stage in order to pass the signal without distortion or loss of power. There are many techniques for matching the output impedance of one section to the input impedance of the next. One of these techniques is to use an impedance-matching transformer.

An impedance-matching transformer works on the same basic principles as power transformers, but more attention is given to the impedance of the primary and secondary windings than is given their voltage and current characteristics.

The impedance of a transfomer winding is determined by a combination of inductance of the coil, resistance of the wire in the coil, the capacitance between the windings, the turns ratio, and the operating frequency.

You already know that a transformer is an inductive device. The primary and secondary windings are coils of wire that behave as inductors. Every transformer winding has a certain amount of inductance that produces an inductive reactance. You have also seen that transformer windings have a small amount of resistance. Finally, there is also a small bit of capacitance between the windings. So a transformer winding is actually an RLC circuit that has an overall impedance that varies with the applied frequency. This fact is normally ignored in the case of power transformers, but it is an essential feature of impedance-matching transformers.

Transformers can be designed so that the impedance of the primary winding is quite different from the impedance of the secondary winding. If there is a need to match the 2000-Ω output of an audio amplifier to an 8-Ω loudspeaker, you want a transformer that has a primary impedance of 2000 Ω and a secondary impedance of 8 Ω. Just as there are mathematical relationships between the turns ratio and the voltage and current transformations, there is a relationship between the turns ratio and impedance transformations:

UNDERSTANDING TRANSFORMERS

$$\frac{Zp}{Zs} = \left(\frac{Np}{Ns}\right)^2$$

where,

Zp is the impedance of the primary winding in ohms,
Zs is the impedance of the secondary winding in ohms,
Np/Ns is the turns ratio (no units).

If you subscribe to a cable-television service, you probably have a small matching transformer connected between the cable and the VHF antenna connection on your television receiver. This transformer matches the 50-Ω impedance of the cable to the 300-Ω impedance of the VHF antenna connections. The primary impedance of this transformer is 50 Ω, and the secondary impedance is 300 Ω. What is the turns ratio?

$$\frac{Zp}{Zs} = \left(\frac{Np}{Ns}\right)^2$$

$$\frac{50}{300} = \left(\frac{Np}{Ns}\right)^2$$

$$0.167 = \left(\frac{Np}{Ns}\right)^2$$

$$\frac{Np}{Ns} = 0.409$$

So a cable-television matching transformer has about 2.5 times as many turns in its secondary as it does in its primary winding. Voltage and current transformations take place, but they are not as important as the impedance-matching effects.

Most impedance-matching transformers are designed to operate over a wide range of frequencies.

Unlike power transformers, impedance-matching transformers are often designed to operate over a wide range of frequencies. This is done by designing matching transformers with a very low-Q factor. Recall that the Q factor of a coil of wire is the ratio of its reactance to its resistance, X_L/R. Using an ohmmeter, you can thus expect to measure relatively high winding resistances. It isn't unusual to see winding resistances of 50 Ω or more. By contrast, high-Q power transformer windings have resistances of less than 1 Ω.

Figure 18-4 shows examples of the two kinds of matching transformers described thus far: an audio transformer and a matching transformer for cable-television hookups. Audio transformers look very much like small power transformers. Audio transformers are rated according to the primary and secondary winding impedances as well as the power (in watts) that they can safely handle. Antenna matching transformers are rated according to their primary and secondary impedances. Their power ratings are not important because the power level is so small—a few milliwatts at most.

**Figure 18-4.
Impedance-Matching
Transformers**

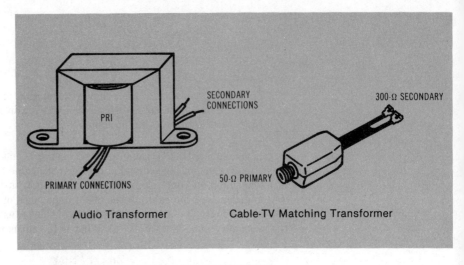

RADIO-FREQUENCY TRANSFORMERS

Radio-frequency (rf) transformers, such as the one illustrated in *Figure 18-5*, are constructed to operate at the much higher frequencies that are commonly called radio frequencies. The range of radio frequencies where such transformers are useful is between 100 kHz (100,000 Hz) and about 10,000 MHz (10,000 million hertz).

**Figure 18-5.
Radio-Frequency
Transformer**

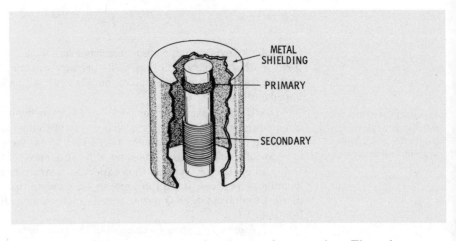

Rf transformers are used as impedance-matching transformers and the inductive portion of LC-tuned circuits in high-frequency equipment.

Rf transformers serve two functions at the same time. First, they serve as impedance-matching transformers between sections in a piece of rf equipment—radio or television receiver, radar, and so on. Second, they serve as the inductive portion of a parallel LC circuit.

Usually the primary and secondary windings of rf transformers can be tuned by turning a ferrite core in or out of the windings. This makes it possible to adjust the inductance of the windings and, therefore, their impedance as well.

UNDERSTANDING TRANSFORMERS

When a circuit is to operate at a very narrow range (or band) of rf frequencies, the primary and secondary windings can be adjusted so that they have a resonant frequency that is equal to the frequency to be passed through them. This produces the narrow-band response shown in *Figure 18-6*.

Figure 18-6.
Tuned Radio-Frequency Transformer Stages

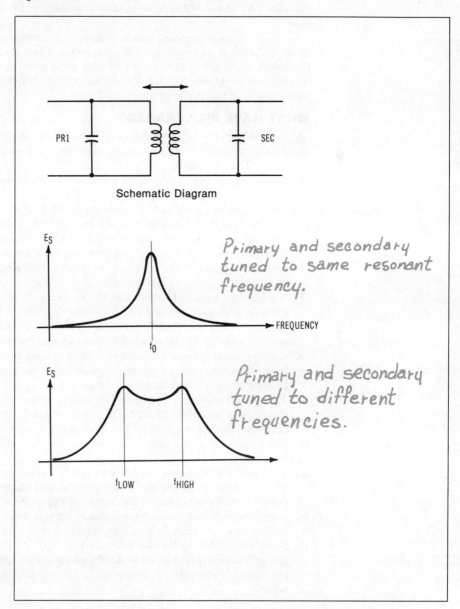

On the other hand, it is sometimes more desirable to pass a wider band of frequencies. When this is the case, the primary winding of the tranformer can be tuned a bit below the desired frequency and the secondary a bit above the desired frequency. Although the transformer is thus detuned and passes less power from the primary to the secondary, it passes a wider band of frequencies than any sharply tuned transformer can pass.

Rf transformers are generally specified according to the number of turns, inductance, and resistance of their primary and secondary windings. Engineers can use that information to determine whether they can use the transformer to get the frequency response, Q factor, and overall impedance they need for a particular application.

WHAT HAVE WE LEARNED?

1. A transformer is an inductive device that has at least two windings.
2. A basic transformer has a primary winding and a secondary winding wrapped around the same core.
3. Power applied to the primary winding of a transformer is magnetically induced into the secondary winding.
4. The amount of power at the secondary winding of a transformer cannot be greater than the amount of power being applied to the primary winding.
5. The purpose of a power transformer is to alter voltage and current levels to match the requirements for operating electrical and electronic equipment.
6. Power transformers are designed to operate from sine waveforms at one particular frequency (usually 60 Hz).
7. A step-down transformer reduces the ac voltage level applied to its primary winding. A step-up transformer increases the voltage level.
8. The turns ratio of a transformer is the ratio of the number of turns in the primary winding to the number in the secondary winding.
9. A step-up transformer has a turns ratio less than 1 (more secondary turns than primary turns). A step-down transformer has a turns ratio greater than 1 (more primary turns that secondary turns).
10. A step-up transformer increases the applied voltage but decreases the current a proportional amount. On the other hand, a step-down transformer decreases the applied voltage but increases the current.
11. The power rating of a power transformer is specified in watts or volt-amperes.
12. Most power transformers are more than 97% efficient. The losses are due to the resistance of the wiring, eddy currents, and magnetic hysteresis.
13. Power transformers tend to run warm or hot to the touch.
14. Many power transformers have a center-tapped secondary winding for special applications, and some have a center-tapped primary winding.
15. Impedance-matching transformers work according to the same principles as power transformers, but serve a different purpose—matching the output impedance of one section of a circuit to the input impedance of the next section.

UNDERSTANDING TRANSFORMERS

16. The impedance of a transformer winding is determined by a combination of inductance of the coil, resistance of the wire in the coil, the capacitance between the windings, and the operating frequency.
17. The turns ratio of a transformer determines its impedance ratio.
18. Most impedance-matching transformers are designed to operate over a wide range of frequencies.
19. Rf transformers are used as impedance-matching transformers and the inductive portion of LC-tuned circuits in high-frequency equipment.
20. The inductance value of many rf transformers can be adjusted by turning a ferrite core in and out of the coil.

KEY WORDS

Audio transformer
Impedance-matching transformer
Impedance ratio
Isolation transformer
Primary winding
Rf transformer
Secondary winding
Turns ratio

18 UNDERSTANDING TRANSFORMERS

Quiz for Chapter 18

1. Which one of the following statements is true?
   a. The secondary voltage for a step-down power transformer is greater than the primary voltage.
   b. The secondary current for a step-down power transformer is greater than the primary current.
   c. Power at the secondary winding of a step-down transformer is greater than the power applied to the primary.
   d. A step-down transformer has a turns ratio less than 1.

2. The secondary voltage for a certain power transformer is twice the voltage applied to the primary. The current flowing in the secondary winding will be:
   a. One-half the primary current.
   b. Same as the primary current.
   c. No more than 3% less than the primary current.
   d. Twice the primary current.

3. What is the turns ratio of a power transformer that shows 50 V at its secondary winding when you apply 150 V to its primary winding?
   a. 1/3.
   b. 1/2.
   c. 2.
   d. 3.

4. A certain power transformer has a turns ratio of 5. What voltage can you expect at the secondary winding when you apply 120 V to the primary winding?
   a. 0.417 V.
   b. 6 V.
   c. 24 V.
   d. 600 V.

5. A power transformer steps 100 V down to 10 V. If the current in its primary winding is 2 A, what is the current in the secondary winding?
   a. 0.2 A.
   b. 2 A.
   c. 5 A.
   d. 20 A.
   e. 50 A.
   f. 100 A.

6. Assuming 100% efficiency, what is the power being applied to the primary winding of the transformer described in Question 5?
   a. 2 W.
   b. 10 W.
   c. 20 W.
   d. 100 W.
   e. 200 W.
   f. 500 W.

7. Assuming 100% efficiency, what is the power being consumed at the secondary winding of the transformer described in Question 5?
   a. 2 W.
   b. 10 W.
   c. 20 W.
   d. 100 W.
   e. 200 W.
   f. 500 W.

UNDERSTANDING TRANSFORMERS

8. An audio transformer has a turns ratio of 9. What is its impedance ratio?
   a. 3.
   b. 4.5.
   c. 6.
   d. 9.
   e. 18.
   f. 81.

9. The primary impedance of a transformer is 60 Ω and the secondary impedance is 120 Ω. What is the impedance ratio?
   a. 0.25.
   b. 0.707.
   c. 0.5.
   d. 2.
   e. 4.
   f. 180.

10. What is the turns ratio for the transformer specified in Question 9?
    a. 0.25.
    b. 0.707.
    c. 0.5.
    d. 2.
    e. 4.
    f. 180.

Glossary

The definitions in this glossary apply to the terms as they are used in this book. Consult a dictionary of electronics terms for more general definitions.

AND Logic: An arrangement of two or more switches connected in series in such a way that current flows through the circuit only when all switches are closed. Compare with "OR Logic."

Active Element: A chemical element that requires little external force to free the electrons in the outer orbits of its atoms. Good conductors of electricity are usually made of active elements.

Alternating Current (ac): Current through a circuit that is characterized by changing direction, or polarity, at regular intervals. Compare with "Direct Current (dc)."

Ammeter: An electrical instrument used for measuring the amount of current flowing in a circuit.

Ampere (A): The basic unit of measurement for current.

Ampere-Turns: A measure of an electromagnet's strength, expressed in terms of the amount of current flowing through the electromagnet multiplied by the number of turns of wire.

Amplitude: The level of current or voltage in a circuit.

Armature: The rotating portion of a basic motor or generator.

Atom: The smallest possible particle of an element that retains the character of that element. Atoms are composed of electrons, protons, and neutrons.

Audio Transformer: An impedance-matching transformer that is specifically designed to match the impedances between two sections of audio circuits.

Bar Magnet: A permanent magnet that has the shape of a straight bar or rod. Compare with "Horseshoe Magnet."

Baseline: The portion of a pulse waveform that is taken as the point of reference for the minimum voltage or current level.

Battery: Through common usage of the term, a battery is an electrical device that converts chemical energy to electrical energy. A more strict definition refers to a device that is made up of two or more cells.

Capacitive Reactance: The opposition to ac current flow created by a capacitor. The basic unit of measurement is the ohm. Compare with "Inductive Reactance."

GLOSSARY

Capacitor: An electrostatic device that is composed of at least two conductive plates separated by a dielectric material.

Carbon-Zinc Cell: A primary cell that has electrodes made of carbon and zinc. The standard voltage is about 1.5 V.

Cell: A basic device that converts one form of energy to dc current and voltage. Examples are chemical cells (chemical energy to electrical energy) and solar cells (light energy to chemical energy).

Centrifugal Switch: A switch that is opened or closed by centrifugal force. These switches are attached to spinning portions of machinery and are often used to indicate whether or not the spinning action is actually taking place.

Choke: See "Inductor."

Coil: (1) The turns of wire in an electromagnetic device. (2) An alternative name for inductor.

Combination Circuit: A circuit that has some components connected in series and others in parallel.

Commutator: The portion of a motor or generator where fixed electrical connections are made to the movable armature section.

Conductor: Any material that passes electrical current with relative ease. Compare with "Insulator."

Counter Emf: The reverse-polarity voltage (or emf) that is produced by a changing current level applied to an inductive device. Counter emf is reponsible for the fact that current through an inductor lags the applied voltage by 90°.

Current: The flow of electrons through a conductor. The basic unit of measurement is the ampere.

Current Divider: A parallel circuit that divides the applied current into two or more paths. Also see "Shunt Resistor."

Cycle: For a sine waveform, a cycle is one complete change from zero to the positive peak value, back through zero to the negative peak value, and back to zero again.

Dc Generator: An electromechanical device that converts mechanical motion (usually rotary motion) into dc voltage.

Decay Time: The time required for a pulse waveform to fall from 90% of its peak value to 10% of its peak value. Compare with "Rise Time."

Diamagnetic Material: A material that cannot be magnetized under normal circumstances. Compare with "Paramagnetic" and "Ferromagnetic" materials.

Dielectric: An insulating material commonly used for separating the conductive plates of a capacitor.

Dielectric Constant: A measurement of how well a dielectric material concentrates electrostatic lines of force, where air has a dielectric constant of 1. This value has no unit of measurement.

Dielectric Strength: A measurement of how much voltage a dielectric material can withstand without breaking down. It is usually expressed in terms of volts per mil (volts per 0.001 inch).

Direct Current (dc): Current through a circuit that is characterized by flowing only in one direction. Compare with "Alternating Current (ac)."

Dry-Cell Battery: A battery that uses a dry or paste-like electrolyte. Compare with "Wet-Cell Battery."

Electrolyte: A good conductor that passes current in the form of ions as opposed to electrons.

Electrolytic Capacitor: A type of capacitor that uses a polarized electrolyte to create a very thin dielectric barrier. Such capacitors are intended for dc applications.

Electromagnet: A magnet that is created by current flowing through a coil of wire. It is usually a temporary magnet that has a magnetic field only as long as current is flowing through the coil.

Electromotive Force (emf): Electrical force; often called "voltage."

Electrons: Tiny, negatively charged particles that surround the nucleus of atoms. Electrons that are freed from their parent atoms can move through a good conductor as current.

Electrostatic Lines of Force: Basic units of force created by electrical charges.

Element: A basic chemical substance that forms the building block for molecules.

Emf: See "Electromotive Force (emf)."

Equivalent Resistance: A single resistance value that represents a combination of two or more separate resistance values. Equivalent resistances are often used to simplify the analysis of circuits.

Farad (F): The basic unit of measurement for capacitance. One farad is the amount of capacitance required for charging the plates at 1 V per second when the charge current is 1 A.

Ferromagnetic Material: A material that is easily magnetized. Compare with "Diamagnetic" and "Paramagnetic" materials.

Flux density: A measure of the concentration of lines of force in a magnetic field. The higher the flux density, the stronger the magnetic effect.

Flux Lines: The unit lines of force for magnetic fields.

GLOSSARY

Free Electron: An electron that has been freed from its parent electron.

Frequency: The number of ac cycles that are completed in one second. The basic unit of measurement is the hertz (Hz), where 1 Hz is one complete cycle per second.

Generator: An electrical mechanical device that converts rotary mechanical motion into electrical energy.

Ground Reference: A common point of reference for measuring voltages in a circuit.

Henry (H): The unit of measurement for inductance. One H of inductance is the amount required to generate 1 V of counter emf when current through the inductor is changing at the rate of 1 A per second.

Hertz (Hz): The basic unit of measurement for frequency.

High-Pass Filter: An electronic circuit that tends to pass higher frequencies better than lower frequencies. Compare with "Low-Pass Filter."

Horseshoe Magnet: A permanent magnet that has the shape of a horseshoe. Compare with "Bar Magnet."

Hysteresis: A characteristic of a magnetic core material where changes in the magnetic field lag the corresponding change in magnetizing current.

Hysteresis Loss: Electrical energy expended to overcome hysteresis in the core material of an inductive device.

IR Voltage Drop: The voltage across a resistive device as determined by the product of the value of the resistance times the amount of current flowing through it.

Impedance: The opposition to ac current flow offered by a combination of resistance and reactance (inductive or capacitive or both). The basic unit of measurement is the ohm.

Impedance Ratio: The ratio of the primary to the secondary impedance of a transformer.

Impedance-Matching Transformer: A transformer specifically designed to match the impedance of one stage of a circuit to the impedance of the next.

Inductance: The property of a device that opposes a change in current flow in response to a change in applied voltage. The basic unit of inductance is the henry.

Induction: The process that is reponsible for electrical and magnetic effects taking place at a distance through electrostatic and magnetic fields.

Inductive Reactance: The opposition to ac current flow caused by an inductance. The basic unit of measurement for inductive reactance is the ohm. Compare with "Capacitive Reactance."

GLOSSARY

Inert Element: An element that does not readily give up free electrons from the outer orbits of its atoms.

Insulator: A material that does not conduct electrons under normal operating conditions. Compare with "Conductor."

Ion: A molecule of a substance that has an abnormally small or large number of electrons. Negative ions have an excess of electrons, while positive ions have a shortage of electrons.

Isolation Transformer: A transformer that is designed to isolate current flow in one winding from the current flow in the other.

Kilohm (kΩ): One thousand ohms.

Kilovolt (kV): One thousand volts.

Kilowatt (kW): One thousand watts.

Knife Switch: A primitive, manually operated switch that uses a metal blade to make and break contact with a terminal.

Lagging Waveform: An ac current or voltage waveform that is delayed relative to a reference waveform. Compare with "Leading Waveform."

Lead-Acid Cell: A secondary wet cell that uses electrodes made of lead compounds and an acid electrolyte. Auto batteries are made of lead-acid wet cells.

Leading Edge: The edge of a pulse waveform that immediately follows the baseline level. Compare with "Trailing Edge."

Leading Waveform: An ac current or voltage waveform that is ahead of a reference waveform. Compare with "Lagging Waveform."

Left-Hand Rule: A technique for determining the relative directions and polarities of currents and magnetic fields; it uses the thumb, index, and middle fingers of the left hand as "pointers."

Loadstone: A naturally occurring material that exhibits the properties of a permanent magnet.

Low-Pass Filter: An electronic circuit that tends to pass low frequencies better than high frequencies. Compare with "High-Pass Filter."

Magnet: Any device or material that exhibits the properties of magnetism: deflecting a compass needle, for example.

Magnetic Field: The invisible force field that surrounds a magnet.

Magnetic Lines of Force: The unit lines of force that make up a magnetic field.

Magnetic Poles: The starting and ending points for the magnetic lines of force for a magnet. The poles are labeled north (N) and south (S).

Magnetic Shielding: A material that deflects magnetic lines of force in order to avoid unwanted magnetic interaction with another device.

Magnetism: The essential property of magnets; characterized by a force that interacts with other magnets and ferromagnetic materials.

Magnetomotive Force (mmf): The force that creates and maintains magnetic lines of force around a magnet or electromagnet. It is the magnetic counterpart of voltage.

Maxwell: A basic unit of measurement of magnetic flux. (Also see "Weber.")

Megohm (MΩ): One million ohms.

Mercury Cell: A rechargeable, secondary cell that has a mercuric-oxide cathode. The standard voltage is about 1.35 V.

Microampere (μA): One-millionth of an ampere.

Microfarad (μF): One-millionth of a farad.

Microvolt (μV): One-millionth of a volt.

Milliampere (mA): One-thousandth of an ampere.

Millivolt (mV): One-thousandth of a volt.

Mmf: See "Magnetomotive Force."

Molecule: The smallest particle of a substance that retains the physical and chemical characteristics of that substance.

Motor: Any device that changes electrical energy into mechanical force and motion. This term is most often applied to devices that convert electrical power into rotary motion.

Neutrons: Particles within the nucleus of an atom that have no electrical charge. Usually the number of neutrons in an atom is equal to the number of protons.

Nickel-Cadmium (Nicad) Cell: A rechargeable, secondary voltage cell that has electrodes of nickel and cadmium. The cell potential is about 1.2 V.

Normally Closed Switch: A switch that uses a spring to hold the contacts closed. Depressing the switch opens these contacts.

Normally Open Switch: A switch that uses a spring to hold the contacts open. Depressing the switch closes these contacts.

Ohm: The basic unit of measurement for resistance.

Ohm's Law: The fundamental expression for the relationship between current, voltage, and resistance.

Ohmmeter: An electrical instrument that measures resistance.

GLOSSARY

Oscillator: An electronic circuit that converts a dc voltage to an ac voltage or a pulsating dc voltage.

Parallel Circuit: A circuit that provides more than one path for current. Compare with "Series Circuit."

Paramagnetic Material: A material that can be magnetized only with great difficulty. Compare with "Diamagnetic" and "Ferromagnetic" materials.

Peak Amplitude: The highest level reached by a voltage or current waveform.

Period: The time required to complete one full cycle of a waveform.

Permanent Magnet: A magnet that retains its magnetic field long after the magnetizing force is removed. Compare with "Temporary Magnet."

Permeability: A characteristic of a material that indicates how easily it can take on a magnetic field.

Permeance: A measure of how easily a material can conduct a magnetic field. Compare with "Reluctance."

Picofarad (pF): One-millionth of a microfarad.

Piezoelectric Effect: A property of certain materials to generate a dc voltage when mechanical pressure is applied to them and to undergo mechanical stress when a voltage is applied.

Potential Difference: The difference in voltage potential between two points in a circuit.

Power: The rate of doing work. In electronics, this means the rate of converting electrical energy to some other form of energy. Resistors consume power by converting electrical energy to heat. The basic unit of measurement for power is the watt (W).

Primary Cell: A chemical cell that does not require charging before it can be used. Primary cells are not rechargeable.

Primary Winding: The winding of a transformer where the voltage, current, or impedance to be "transformed" are applied. Compare with "Secondary Winding."

Product-Over-Sum Rule: A procedure for determining the total resistance of two resistances connected in parallel and the total capacitance of two capacitors connected in series. The procedure is to divide the product of the two values by their sum.

Protons: Particles within the nucleus of an atom that have a positive electrical charge. Usually the number of protons in an atom is equal to the number of neutrons.

Pulsating Dc: A type of waveform characterized by a dc voltage that is switched on and off at regular intervals.

GLOSSARY

Pushbutton Switch: A switch that is opened or closed by depressing a pushbutton. (See "Normally Open Switch" and "Normally Closed Switch" for examples.)

Rf Transformer: A transformer that is specifically designed to operate at the high ac frequencies such as those encountered in radio, television, and radar equipment.

Reluctance: A measure of a material's opposition to the conduction of magnetic fields. It is the magnetic counterpart of resistance to current flow. Compare with "permeance."

Residual Magnetism: The amount of magnetism that remains in a core of an electromagnet when the current through the coil is reduced to zero.

Resistance: The opposition to current flow offered by resistive materials that dissipate energy in the form of heat. The basic unit of resistance is the ohm.

Resistor: An electronic component specifically designed to offer a certain amount of resistance to current flow.

Resonant Frequency: The unique frequency in series and parallel LC circuits where the inductive and capacitive reactances are equal.

Retentivity: The ability of a material to retain its magnetic field after the magnetizing force is removed.

Rise Time: The time required for a pulse waveform to rise from 10% of its peak value to 90% of its peak value. Compare with "Decay Time."

Root-Mean-Square (rms): The value of ac voltage or current required to do the same amount of work as the same value of dc voltage or current. For sine waveforms, this value is equal to 0.707 times the peak value.

Saturation: A characteristic of electromagnets where increasing the amount of current flowing through the coil of wire causes no further increase in the strength of the magnetic field.

Secondary Cell: A chemical cell that requires (and permits) recharging. An automobile battery, for example, is made up of secondary cells.

Secondary Winding: The winding on a transformer that provides the "transformed" values of voltage, current, and impedance. Compare with the "Primary Winding."

Selector Switch: A switch that is designed so that you can select one of any number of settings.

Series-Aiding: A series arrangement of voltage sources where the positive pole of one source is connected to the negative pole of the other. The total voltage is equal to the sum of two source voltages. Compare with "Series-Opposing."

GLOSSARY

Series Circuit: A circuit that provides only one path for current. Compare with "Parallel Circuit."

Series-Opposing: A series arrangement of voltage sources where the positive pole of one source is connected to the positive pole of the other. The total voltage is equal to the difference between the two source voltages. Compare with "Series-Aiding."

Shunt Resistor: A resistor connected in parallel with a meter movement in order to reduce the amount of current flowing through the movement.

Sine Waveform: An ac waveform that is characterized by the vertical displacement of a rotating voltage vector. It is the most common type of ac waveform.

Solar Cell: A device that converts light energy to a dc electrical potential.

Solenoid: An electromagnetic device that uses an iron core that is free to move in response to a magnetizing current flowing through a coil of wire.

Static Electricity: Electricity that is generated by friction between two unlike materials.

Temporary Magnet: A magnet that loses its magnetic character the moment the magnetizing force is removed. Most temporary magnets are electromagnets.

Thermocouple: A device that changes a difference in temperature between two points into a dc voltage.

Time Constant: An indication of the amount of time required for dc current (RL circuit) or dc voltage (RC circuit) to rise to 63.2% of its peak value or fall to 63.2% of its peak value.

Toggle Switch: A spring-loaded switch that can be set to its open or closed position. A common wall switch is an example of a toggle switch.

Toroid: A donut-shaped core for inductors and transformers. This shape confines the magnetic field to the core material.

Trailing Edge: The edge of a pulse waveform that follows the peak value and returns to the baseline level. Compare with "Leading Edge."

Transformer: A device that transforms voltage, current, and impedence levels through the interactions of magnetic fields between the primary winding and the secondary winding.

Turns Ratio: The ratio of the number of turns of wire in the primary winding to the number in the secondary winding of a transformer.

Vector: A mathematical "arrow" whose length indicates an amount of current or voltage and whose angle represents an amount of phase shift.

Volt (V): The basic unit of measurement for voltage.

Voltage: The force that causes current to flow through a conductor. The basic unit of measurement is the volt. (Also see "Electromotive Force (emf).")

Voltage Divider: A series arrangement of passive components (resistors, inductors, or capacitors) that is used for dividing a source voltage into smaller portions.

Voltage Drop: The difference in voltage potential between two points in a circuit. (Also see "IR Voltage Drop.")

Voltmeter: An electrical instrument that is used for measuring voltage.

Watt (W): The basic unit of measurement for electrical power.

Waveform: A graphical representation of how voltage, current, or power changes level over time.

Weber: A basic unit of measurement of magnetic flux. (Also see "Maxwell.")

Wet-Cell Battery: A chemical-cell battery that uses a liquid electrolyte. Most wet-cell batteries are rechargeable, secondary-cell batteries.

Index

Ac Circuits
   analyzing combination, 152-153
   analyzing parallel, 149-151
   analyzing series, 146-149
   basic, 145-146
Ac current flow, 131-133
Ac sources, 132-133
Active element, 5
Ac voltage and current, effect of capacitance on, 224-225
Ac waveforms, 134
Aging, 109
Alternator, 29
Ammeter, 20, 21
Amperes, 20
Ampere-turns, 117
AND circuit, 66
Antenna-matching transformers, 297
Arctangent functions, 195
Armature, 124
Atomic particles, 4
Atoms, 3
   atomic particles, 4
   free electrons, 4-5
   inert elements, 5
   ions, 5-6
Audio transformers, 297

Battery, 43
   use of, in parallel circuits, 80
Battery chargers, 27
Branch, 71
Branch currents
   in parallel capacitor circuit, 238-240
   in parallel inductor circuit, 180

Capacitance, 215-220
   effect of, on Ac voltage and current, 224-225
   effect of, on pulse waveforms, 226
   stray, 226-227
   total
      in parallel capacitor circuit, 237
      in series capacitor circuit, 241-242

Capacitive circuit, 231-232
   application of Dc power equations to, 235
   application of Ohm's law to, 232-235
Capacitive reactance, 225-226
   in parallel capacitor circuit, 237-238
   in RLC circuit, 278
   in series capacitor circuit, 242-243
Capacitor
   charging, 218
   connected in parallel, *See* Parallel capacitor circuit
   connected in series. *See* Series capacitor circuit
   construction of, 221-222
   determining value of, 222-223
   discharging, 219
   rating of, 220-221
Carbon-zinc batteries, 24-25
Carbon-zinc cell, 24
Center-tapped transformers, 295-296
Centrifugal switch, 98
Chemical voltage sources, 24-27
   dry cells, 24-25
   wet cells, 25-27
Chlorine, 6
Choke, 157
Coil, 157
Combination ac circuits, analyzing, 152-153
Combination circuits, 87-89
   total current, 91-93
   total resistance, 89-91
   total voltage, 91-93
Combination inductive circuits, analysis of, 180-184
Commutator, 124
Compensating magnets, 108
Conductors, 7-9
Consequent poles, 107
Conventional current flow, 19
Copper atom, 5
Counter emf, 157, 158
Current, 19
   amperes, 20
   common values for, 22-23
   current flow theory, 19-20
   definition of, 1

INDEX

in parallel RC circuit, 261-263
in parallel RLC circuit, 282-284
for a power transformer, 292-293
in RLC circuits, 278
total
    in combination circuit, 91-93
    in series circuit, 60
    in series RC circuit, 257-258
    in series RLC circuit, 281-282
    in series RL circuit, 198
    use of Ohm's law for solving for, 49
Current flow, 72-74
    total opposition to, in series RL circuit, 195-197
Current flow theory, 19-20

Dc current flow, 131
Dc generator, 43
Dc parallel circuits, 71-81
Dc power equations
    application of
        to capacitive circuits, 235
        to inductive circuit, 172-174
Dc series circuits, 57-66
Dc waveforms, 133-134
Decay time, 141
Demagnetizing metals, 106-107
Dielectric, 215
Dielectric constant, 223
Dielectric strength, 223
Digital computers, 140
Dry cells, 24-25

Eddy current, 294
Electrical charges, 10-13
Electrical circuit, 43
Electrical potential, 20
Electricity, definition of, 1-2
Electric power, 51
    power equations, 51-53
Electric thermometer, 29
Electrodes, 24
Electrolyte, 24
Electrolytic capacitors, 221
Electromagnetism, 115
    applications of, 119-121
    hysteresis, 118-119

in motors, generators, and meters, 121-126
polarity, 117-118
residual magnetism, 118
strength, 116-117
Electromagnets, polarized, 120-121
Electromotive force, 20
Electron(s), 3
    charge of, 4
    definition of, 1
    free, 4-5
Electron flow, and energy, 7-9
Electron flow theory, 20
Electrostatic lines of force, 13-14
Elements, 3
    active, 5
    inert, 5
Energy, and electron flow, 7-9

Fixed resistor, 35, 64
Fluorine, 5
Flux density, 110
Flux loops, 109
Free electrons, 4-5

Generators, electromagnetism in, 124-125
Ground, 63

Heat-generated voltage, 29-30
Helium, 4
Henry, 157
High-pass filter, 163
Hysteresis, 118-119

Impedance, 196
    in parallel RC circuit, 263-265
    in parallel RLC circuit, 285
    in RLC circuit, 278
    in series RC circuit, 256-257
    in series RLC circuit, 279-280
Impedance-matching transformers, 289
    principles of, 296-298
Inductance, 157-158
    applications of, 163-166
    effect of, on ac current, 159-161
    total
        in parallel inductor circuit, 178

INDEX

in series inductor circuit, 174
value of, 161-162
Induction, principle of, 12
Inductive circuits, 169-170
   application of
      Dc power equations to, 172-174
      Ohm's law to, 170-172
Inductive reactance, 162-163, 169
   in parallel inductor circuit, 178-179
   in RLC circuit, 276, 278
   in series inductor circuit, 175
Inductor, 157
   Q factor of, 199-200
   connected in parallel. *See* Parallel inductor circuit
   connected in series. *See* Series inductor circuit
Inert elements, 5
Insulators, 7-9
Ions, 5-6
IR drop, 51
Isolation transformer, 292

Keeper, 108
Kilohertz, 140
Kirchhoff's current law, 95-97
Kirchhoff's law, 94
Kirchhoff's voltage law, 94-95
Knife switches, 45

Lead-acid battery, 26
Lead-acid cell, 24
Left-hand rule, 29, 125
Light-generated voltages, 31-32
Lightning, 9, 19
Loadstone, 101, 102, 108
Low-pass filter, 163

Magnetic field, 104-106
Magnetic lines of force, 104
Magnetic poles, 107
Magnetic shielding, 106
Magnetic voltage sources, 28-29
Magnetism, 101-102
   basic units of, 109-111
   demagnetizing metals, 106-107
   magnetic characteristics, 102
   magnetic field, 104-106

magnetic poles, 107
molecular theory of, 103-104
permanent magnets, 108-109
residual, 118
temporary magnets, 108-109
Magnetomotive force (mmf), 109-110
Magnets, applications of, 102
Maxwell, 110
Megahertz, 140
Mercury batteries, 25
Mercury cell, 24
Metals, demagnetizing, 106-107
Meters, electromagnetism in, 125-126
Microampere, 20
Microphones, 32
Milliammeter, 29
Milliampere, 20, 29
Molecular theory of magnetism, 103-104
Molecules, 2-3
Momentary-pushbutton switches, 46
Motors, electromagnetism in, 123-124

Negative point, 2
Neutrons, 3
   charge of, 4
Nicad batteries, 27

Oersted, Hans C., 115
Ohmmeter, 37, 165
Ohms, 34
Ohm's law, 48
   applications of
      to ac circuits, 145
      to capacitive circuits, 232-235
      to inductive circuits, 170-172
   solving for
      current, 49, 60, 92, 147, 159
      resistance, 49-50, 65, 74, 150
      voltage, 49
Oscillator, 133

Parallel ac circuits, analyzing, 149-151
Parallel capacitor circuit, 236-237
   determining branch currents in, 238-240
   determining capacitive reactances in, 237-238

determining total capacitance, 237
Parallel circuits, 71-72
  current dividers, 78-79
  current flow, 72-74
  total resistance, 74-77
  use of batteries in, 80
  use of switches in, 80-81
Parallel inductor circuit, 176-178
  determining branch currents in, 180
  determining inductive reactances in, 178-179
  determining total inductance, 178
Parallel RC circuit
  analysis of, 265-266
  current in, 261-263
  impedance of a, 263-265
  voltage in, 260-261
Parallel RLC circuit
  currents in a, 282-284
  effect of frequency on, 282-285
  impedance of a, 285
  resonant frequency of a, 284-285
Parallel RL circuits, 200
  analysis of, 206
  current in, 201-204
  impedance of, 205-206
  voltage in, 200-201
Permanent magnets, 108-109
Permeance, 110-111
  levels of, 111
Photodiodes, 31
Phototransistors, 31
Piezoelectric materials, 33
Polarity, 2, 117-118
Polarized electromagnets, 120-121
Positive point, 2
Potassium hydroxide, 27
Potential difference, 20
Power, 51
  determination of, 51
Power equations, 51-53
Power transformers, 289
  currents for, 292-293
  efficiency and losses, 294-295
  principles of, 290-296
  ratings of, 294
  turns ratio of, 291
  voltages for, 291-292

Pressure-generated voltages, 32-33
Primary-cell batteries, 25
Primary winding, 289
Product-over-sum rule, 75-77, 153
Protons, 3
  charge of, 4
Pulse waveforms
  effect of capacitance on, 226
  effect of RC circuits on, 266-268
  effect of RL circuits on, 206-209
  measuring, 140
  rise and decay time, 141

Q factor
  of an inductor, 199-200
  of an RLC circuit, 285-286

Radio-frequency transformers, 298-300
RC circuits, 251
  effect of, on pulse waveforms, 266-268
  parallel, 260-266
  series, 251-260
Reluctance, 109, 110
Residual magnetism, 118
Resistance, 34
  ohms, 34
  resistors, 34-37
  total
    in combination circuits, 89-91
    in parallel circuit, 74-77
    in series circuit, 58
  use of Ohm's law for solving for, 49-50
Resistors, 34-37
  ratings, 36-37
Resonant frequency
  of a parallel RLC circuit, 284-285
  of a series RLC circuit, 280-281
Retentivity, 104
Right-hand rule, 125
Rise time, 141
RLC circuits, 275-276
  current for, 278
  effect of frequency on impedance of R, L, and C devices, 276-278
  impedance for, 278
  Q factor of, 285-286

INDEX

voltage for, 278
RL circuits, 189
   determining Q factor, 199–200
   effect of, on pulse waveforms, 206–209
   parallel, 200–206
   series, 189–199
Rotary switches, 46–47

Sawtooth waveform, 134
Secondary winding, 289
Selector switches, 46–47
Selenium, 31
Series ac circuits, analyzing, 146–149
Series-aiding circuits, 59
Series capacitor circuit, 240–241
   determining voltage drops in, 243–245
   determining capacitive reactances in, 242–243
   determining total capacitance in, 241–242
Series circuit, 57
   total current, 60
   total resistance, 58
   total voltage, 58–59
   use of switches in, 66
   voltage dividers, 63–66
   voltage drop, 61–62
   voltage reference point, 63
Series inductor circuit, 174
   determining inductive reactances in, 175
   determining total inductance, 174
   determining voltage drops in, 175–176
Series-opposing circuits, 59
Series RC circuits, 251–253
   analysis of, 258–260
   impedance in, 256–257
   total current in, 257–258
   voltage in, 253–256
Series RLC circuits
   effect of frequency on, 278–282
   impedance of, 279–280
   resonant frequency of, 280–281
   total current in, 281–282

Series RL circuits, 189–191
   analysis of, 198–199
   total current in, 198
   total opposition to current flow in, 195–197
   voltage in, 191–195
Shunt resistor, 79
Silver atom, 5
Sine wave, 134
   generating, 135–136
   measuring, 136–137
   period and frequency, 139–140
   rms value, 137–138
Sine-wave cycle, 138–139
Sodium, 6
Solar cells, 31
Solenoids, 119
Square-wave pulses, 140
Static electricity, 9–10
   electrical charges, 10–13
   force of charge, 13–14
Step-down transformer, 291
Step-up transformer, 291
Stray capacitance, 226–227
Subatomic particles, 3
Sulfuric acid, 26
Switches, 43, 45
   knife switches, 45
   momentary-pushbutton switches, 46
   rotary switches, 46–47
   toggle switch, 45–46
   use of, in parallel circuits, 80–81
   use of, in series circuits, 66

Temporary magnets, 108–109, 115
Thermocouple generator, 29–30
Toggle switch, 45–46
Tolerance, 37
Toroids, 119–120
Total current
   in combination circuits, 91–93
   in series circuit, 60
   in series RC circuit, 257–258
   in series RLC circuit, 281–282
   in series RL current, 198
Total inductance
   in parallel inductor circuit, 178
   in series inductor circuit, 174

INDEX

Total resistance
   in combination circuit, 89–91
   in parallel circuit, 74–77
   in series circuit, 58
Total voltage
   in combination circuits, 91-93
   in series circuit, 58–59
Transformers, 289
   impedance-matching, 289, 296–298
   power, 289, 290–296
   radio-frequency, 298–300
Turns ratio, of a transformer, 291
Tweeters, 164

Variable resistor, 35
Voltage, 20
   common values for, 22–23
   definition of, 1
   in parallel RC circuit, 260–261
   for a power transformer, 291–292
   in RLC circuit, 278
   in series RC circuit, 253–256
   in series RL circuit, 191–195
   source of, 2, 23–33
   total
      in combination circuits, 91-93
      in series circuit, 58–59
   use of Ohm's law for solving for, 49
Voltage dividers, 63–66
Voltage drop, 50–51, 61–62
   amount of, 61–62
   in series capacitor circuit, 243–245
   in series inductor circuit, 175–176
   polarity of, 62
Voltage reference point, 63
Voltmeter, 21
Volts, 21

Watt, 51
Waveforms, 133
   Ac waveforms, 134
   Dc waveforms, 133–134
   generating sine waves, 135–136
   measuring pulse waveforms, 140–141
   measuring sine waves, 136–137
   Rms value, 137–138
   sine-wave cycle, 138–139
   sine wave period and frequency, 139–140
Weber, 110
Wet cells, 25–27
Wirewound resistors, 36
Woofers, 164

Answers to Quizzes

Chapter 1
1. c
2. a
3. b
4. c
5. b
6. a
7. d
8. c
9. a
10. c
11. a
12. c

Chapter 2
1. b
2. c
3. c
4. b
5. c
6. b
7. c
8. b
9. a
10. c
11. b
12. c
13. a
14. e
15. a
16. b
17. e
18. b
19. a
20. a
21. d
22. d
23. b
24. c

Chapter 3
1. b
2. a
3. b
4. c
5. d
6. f
7. a
8. c
9. d
10. a
11. f
12. b
13. b

Chapter 4
1. c
2. a
3. c
4. a
5. e
6. a
7. d
8. a
9. b
10. c
11. d
12. a
13. a
14. a
15. e
16. b

Chapter 5
1. a
2. b
3. b
4. d
5. d
6. d
7. b
8. c
9. d
10. b
11. e
12. e
13. e
14. d
15. c
16. e
17. f
18. b

Chapter 6
1. e
2. b
3. c
4. b
5. b
6. d
7. c
8. d
9. c
10. b
11. b
12. a
13. a

Chapter 7
1. a
2. b
3. c
4. b
5. c
6. a
7. b
8. a
9. c
10. a
11. e

Chapter 8
1. c
2. b
3. a
4. b
5. a
6. e
7. d
8. a
9. b

Chapter 9
1. c
2. c
3. d
4. b
5. d
6. c
7. a
8. d
9. d

Chapter 10
1. c
2. e
3. b
4. b
5. e
6. e
7. b
8. d
9. d

Chapter 11
1. a
2. c
3. b
4. c
5. a
6. e
7. b
8. c
9. a
10. c
11. e
12. a
13. b

Chapter 12
1. b
2. c
3. c
4. d
5. b
6. b

7. c
8. a
9. d
10. a
11. c
12. e
13. d
14. d
15. c
16. d

Chapter 13

1. c
2. a
3. c
4. a
5. c
6. a
7. b
8. a
9. c
10. b
11. e
12. b
13. a
14. d
15. b
16. b
17. d

18. b
19. b

Chapter 14

1. a
2. b
3. b
4. c
5. d
6. c
7. a
8. b
9. a
10. b
11. b
12. b
13. b
14. b
15. b

Chapter 15

1. b
2. b
3. b
4. f
5. a
6. c
7. b

8. a
9. c
10. d
11. c
12. a
13. a
14. c
15. c
16. c

Chapter 16

1. c
2. a
3. c
4. a
5. c
6. b
7. a
8. b
9. c
10. e
11. b
12. e
13. b
14. b
15. d
16. d
17. c

Chapter 17

1. c
2. a
3. b
4. a
5. b
6. e
7. c
8. b
9. a
10. d
11. c
12. a
13. b

Chapter 18

1. b
2. a
3. d
4. c
5. d
6. e
7. e
8. f
9. c
10. b